Intelligent Manufacturing

Intelligent Manufacturing

A Smart Choice...

Abstract

This book contains general information only, and the author or contributors are not, by means of this publication, rendering accounting, business, financial, investment, legal, tax, or other professional advice or services. Your use of the book's contents is not a substitute for such professional advice or services, nor should it be used as a basis for any decision or action that may affect your finances or your business.

Before making any decision or taking any action that may affect your finances or your business, you should consult a qualified professional. Author does not make representation or warranty, express or implied. Your use of the book's contents is solely at your own risk. This book may contain links to third party content, which author does not warrant, endorse, or assume liability for.

Sunil Kumar Wadhwa

Swadhwa2097@gmail.com

First paperback edition 2023

978-1-80227-915-3 (paperback)
978-1-80227-914-6 (eBook)

Acknowledgments

I feel immense pleasure in writing my first book. Inspired by my beloved parents, family, teachers, colleagues, partners and students of all ages and groups. I was fortunate to travel to many countries in last 30+ years during my professional career and learn while working with professionals of Manufacturing, Industrial Automation, and IT systems. My sincere thanks to all who contributed to my work directly or indirectly and look forward to receiving your patronage and blessings in future as well.

I have witnessed and experienced many changes in technology over past several decades, but nothing as compared to Industry 4.0 revolution. The enormity of impact and complexity has moved every pebble in Industrial world. Being in the middle of this exponential change, one of the greatest challenges is sharing the knowledge and learnings as we embrace the new technology. This book is my bit of contribution to be the bridge to connect among technical & industrial community. While I enjoyed writing this book, I hope this learning and sharing of knowledge will continue to grow with your valuable feedbacks and support.

Contents

Table of Figures

List of Tables

Introduction

At times, white smoke can often mean that material is off-gassing moisture and water vapour, meaning the fire is just starting to consume the material. White smoke can also indicate light and flashy fuels such as grass or twigs.

Smoke is made up of particulates, aerosols and gases, and identifying the characteristics of each in a given smoke plume can be helpful when fighting fires. Reading smoke can tell a firefighter what is currently happening with a fire as well as what might happen in the future. One particularly important factor in predicting fire behaviour is the colour of the smoke emitted.

Smoke is the by-product of the fuels it is burning. The colour of the smoke indicates to firefighters the type and density of the fuels involved, all of which give hints as to what the fire might do next.

Thick, black smoke indicates heavy fuels that are not being fully consumed. At times, black smoke can be an indicator that a manmade material is burning such as tyres, vehicles, or a structure. As a rule, the darker the smoke, the more volatile the fire is. Grey smoke can indicate that the fire is slowing down and running out of materials to burn.

Figure 1: Industrial Smoke's color reveals fuel type

We can apply a similar concept to the data coming out of a manufacturing plant. Intelligently analysing the data is like predicting what is burning or cooking inside the manufacturing plant. Data science is the key to capturing data, and validating its accuracy, aggregation, presentation, analyses, and corrective action are the new norms of today's Intelligent Manufacturing. In short, there's no need to continuously watch smoke coming out of a factory to know what fuel is burning but we can use Smart sensors and web-based technology to analyse the nature of smoke and take corrective actions from remote for fuel/ resource consumption, quality, efficiency, and safety.

We are witnessing an unprecedented industrial revolution of a gigantic scale as we are living in a Big Data age. The digital economy has already surpassed the size of the oil economy, and the World Economic Forum pegs it at $100 trillion by 2025. While the industrial revolution was all about factories, the digital revolution is all about platforms, powered by the cloud and enabled by mobility. Our lives, society, work environment, industry, agriculture, communication, global supply chain, economics, trades, and entire geo-political system stand to transform digitally and drastically. The astronomical scale and complexity of this change have already started to bring several surprises for mankind for good or bad, depending on how it unfolds for respective entities. The nature of this revolution is subtle but real, and enormous compared to anything before since the invention of steam engines!

Figure 2: Industrial revolution at gigantic scale

The stakeholders at the receiving end will not be confined simply to a community, a race or a few countries but to the whole of mankind with or without everybody's knowledge.

However, to some technology groups, societies, classes of people, product vendors, business communities, state agencies and various other stakeholders involved in the exploration and implementation of this new limitless technology, there appear to be no boundaries and progress seems unstoppable. This book supports no vendor, association, or group and is being produced only for academic purposes to capture the essence of the paradigm shift in industry enabled by technology and the nature of evolution of new-age Digital Transformation and its impact on the Manufacturing Industry. Topics covered in this book will bring insight into how the Internet and Industry 4.0 revolution started and acknowledges the contributions by several organizations or bodies to enable manufacturers to achieve operational excellence by enabling visibility and intelligence, not just within the plant or across the enterprise but also to the extended supply chain. As it could be disruptive, if not destructive, for Industry, Processes, Technologies and People, current and future Industry Trends show that a tremendous investment in Digital Transformation is already a part of their business strategy as stated below by a leading research firm IDC (Predictions, 2022):

> *"Direct digital transformation (DX) investment growth rates are accelerating, up to 16.5% CAGR for 2022-2024. This is an increase of one full percentage point from the 15.5% CAGR we saw for the 2020-2023-time frame. Investment levels for 2022-2024 are expected to be $6.3 trillion and are 55% of all ICT investment by the end of 2024. This is a $0.9 trillion increase from the 2020-2023-time frame where we forecast direct investment at $5.4 trillion. For the first time ever, we see that the majority of enterprise organizations (at 53%) have an enterprise-wide digital transformation strategy, a 42% increase from just two years ago," states Shawn Fitzgerald, research director, Worldwide Digital Transformation Strategies.*[1]

Customers pursuing Application Modernization and Digital Transformation were enabled to generate remarkable benefits such as:

- 30% TCO Reduction
- 30% faster time to market
- 50% increase in developer productivity
- Requires 70% fewer resources

The Internet has opened up vast amounts of data. Several companies have designed tools to access that data and make it widely available.

Since it is a man-made revolution and involves many new sub-technologies working in tandem, comprehension and control of it by one set or group of people is impossible. The rate at which the change is unfolding is found to be incomprehensible compared to the conventional. Whatever the outcome, one thing is confirmed – "a Big Fundamental Change" is inevitable in every aspect of life in the office or at home.

The boost in customer experience and expectations from technology-enabled products and services will mean that many conventional industries will feel the disruptive effect in no time.

Change is Imminent

Thanks to Industry 4.0 or Digital Transformation, whichever way we describe it, the billions of people being connected through mobile and network devices are reshaping industry and our lifestyle forever. Enormous infrastructure is required to handle a huge amount of data at an unprecedented processing speed, and this along with increased storage capacity and analysis/processing of information will unleash boundless opportunities for businesses.

As compared with previous industrial revolutions, fourth revolution have evolved at an exponential rate rather than a linear pace. The tectonic plates of connectivity, productivity, user experience and growth are shifting at enormous speed that reshapes the competitive industrial and social landscape, and redefines the boundaries of manufacturing, prosperity, trade and values.

On April 30, 2015, an article in the Wall Street Journal by Christopher Mims entitled "Data Is Now the New Middle Manager" describes how some start-up companies are substituting data for middle managers. According to the article:

> *"Start-ups are nimbler than they have ever been, thanks to a fundamentally different management structure, one that pushes decision-making out to the periphery of the organization, to the people actually tasked with carrying out the actual business of the company. What makes this relatively flat hierarchy possible is that front-line workers have essentially unlimited access to data that used to be difficult to obtain or required more senior management to interpret."*

The article goes on to elaborate that when databases were very expensive and business intelligence software cost millions of dollars, it made sense to limit access to top managers.[2]

In the era of cloud capitalism, power also shifts to the hands of the user, as people determine how and where they want to consume products and services, rebalancing supply, and demand.

The digitally enabled factory of today looks very different from the leading factory of ten years ago – while manufacturers in 2025 will look very different still. This makes digitalization a requirement and Smart Manufacturing an evolving to-do list. Yet the companies that get it right can harvest the huge – scalable – benefits across the entire manufacturing value chain such as:

- Reduced material losses
- Improving customer service
- Better delivery lead times

- Higher employee satisfaction
- Lower environmental impact

More and more enterprises are embracing mobility tools to be able to serve their customers more flexibly. Cloud-edge computing has changed the world of business for good. More and more consumers on the internet stand to define and get benefited from this digital wave. Digital manufacturing will transform the lifestyle of people and increase their expectations in an unprecedented way, thus starting a new era of competitiveness in industry. It opens up a huge opportunity for the transformation of the manufacturing industry and businesses as a whole.

Industry 4.0 Challenges

Industry 4.0 is the biggest structural change of the past 250 years — a transformation of scale, scope and complexity unlike anything humankind has experienced before. Industry 4.0 is a critical element of the overarching Digital Supply Chain that will define the future across all industries and every aspect of our lives. Adopters of Industrial Internet of Things (IIoT) tech gain competitive advantage and achieve quantifiable benefits from their investment. And yet, the number one challenge organizations face while trying to make the decision is building the business case.

A survey on Top Challenges faced by Industry in deploying IIoT Technology is published in the whitepaper 2016, "IIoT AND BIG DATA ANALYTICS: How Manufacturing System Architecture is Being Transformed" by Littlefield, Matthew of LNS Research in 2015 and shows the results below.[3]

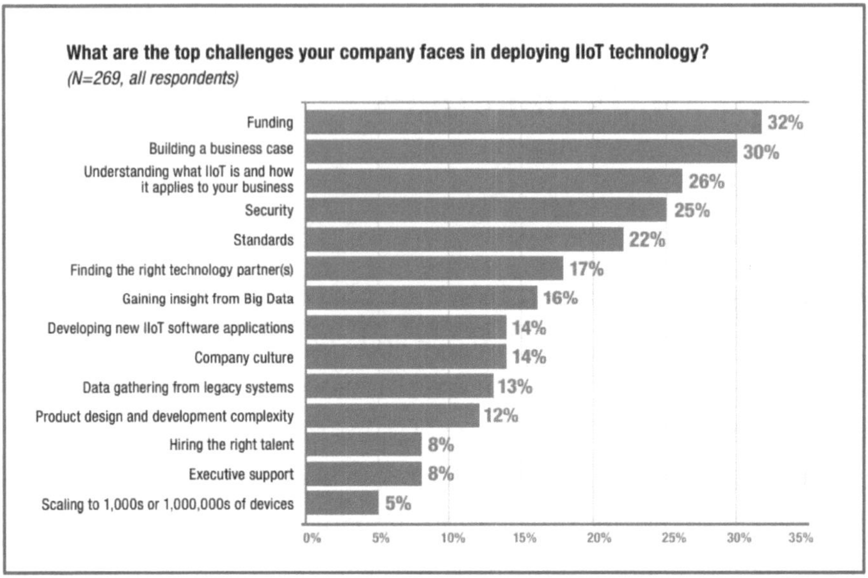

Figure 3: Survey on Top Challenges in Deploying IIoT Technology (Source: LNS Research)

Many existing companies are struggling with fear, uncertainty, a lack of business cases, executive support, funding, and a lack of reliable technology partners and doubt around changes in management of applying new technologies to solve their problems, whereas new players are redefining the industry standards by adopting revolutionary technology. As noted by Tom Kelly, Executive Director & CEO, Automation Alley (2019) in his report "Industry 4.0: From Vision to Implementation:"[4]

> *"The problem lies not in how to get started, but in how to grow teams that can define the problem space, assess the options and understand*

how to gauge value add of any particular solution in terms of metrices that drive operational performance and business value."

Below are the ideas as to what opportunities and threats the manufacturing industry is posing today and how, from the Industry 4.0 (IIoT) revolution, it can gain innovative techniques that could lead to increased benefits and safety measures associated with new technology.

Defined as the merge of our physical, virtual and digital systems into an intelligent system, cyber-physical systems create a tighter integration between humans, machines and information technology systems.

It promises growth and productivity within manufacturing through automation, integration and optimization. The main challenge for manufacturing is that it has been a difficult late adopter – e.g., it was the last to adopt electricity over the steam engine.

"*In my experience, digital transformation initiatives are notoriously difficult to scale up across factory networks – manufacturers may need to slow down to get ahead in the race to implement Industry 4.0,*" says von Scheel.[5]

Billions of dollars are wasted every year due to inefficient and unsafe operations, unexpected plant or equipment downtimes, workman hazards, shutdown, or closure of business. Projected increased output, improved efficiency, enhanced customer experience, product life-cycle improvement, and better safety practices can result in tremendous cost-saving & value addition in industrial operations that can justify diverting the funds for Industry 4.0 implementation to further automate the processes and make them more intelligent – A Smart Move!

Who is Henrik von Scheel?

As Advisory Council member for Federal Minister of Economics & Technology, Henrik inaugurated Germany's national strategy, named the "Digital Agenda". The European Commission adopted it in 2010 as part of the Europe 2020. Today, the "Digital Agenda" is one of Europe's seven flagships – responsible for 5% of GDP, with a market value of €660 billion annually.

Henrik has dedicated his research efforts to 'patternicity' – the science of finding meaning in patterns.

He is best known as the originator of 'Industry 4.0' and the mastermind of the European Digital Revolution which ignited the global themes of today.

His work is applied to national economies, has triggered global themes, influenced GDP growth, reset government policies and shaped the performance of the fastest-growing fortune 100 companies – Apple, Lego, Unilever, ABB, P&G, Siemens and Carlsberg, to name just a few.

Industrial Manufacturing systems are an integration of different technology elements, such as hardware, software, and networks, which can sometimes be complex to configure. The integration of different hardware devices, along with the asset-management software, over the legacy system infrastructure may become complex. Without a strong pre-existing IT ecosystem and skilled personnel, the implementation of the digital infrastructure is a challenge. Manufacturers can do a complete overhaul of the IT system; however, costs incurred in such an overhaul would take a significant amount of time to realize. Even though these changes are imminent, one can take a step-by-step approach to realise the tangible goals with defined outcomes.

Further, legacy systems are often unable to integrate with new-generation smart devices due to protocol issues. These systems are not capable of communicating efficiently with technologically advanced systems. As most manufacturers are still relying on their legacy infrastructure, lack of sensors, and limited communication bandwidth, these integration complexities are hindrances to digital transformation.

As many of us might have felt, Industry 4.0 is coming to our doorstep at such a speed that it is almost impossible to implement the rules and guidelines laid down by authorities and industry associations to minimize the risks and adverse effects on businesses, economies, societies and people. It may not only disrupt conventional businesses and lifestyle but may even transform the livelihoods of many. "Industry 4.0 arrives in waves that disrupt everything"- Manufacturing Leaders' Summit.[6]

Plus, cyber security is yet another challenge for us to tackle as the world embraces Industry 4.0. Further, this book will shed some light on the tangible ways of addressing the following challenges:

1. Journey So far
2. Key-Drivers of Industry 4.0
3. Industry 4.0 Ecosystem
4. Industry 4.0 and Intelligent Manufacturing Initiatives
5. Big Data as fourth Factor of Production
6. Fabric of Industry 4.0
7. Five Layers of Automation Pyramid
8. Design Considerations and Architecture Layers
9. Business Use-Cases

The topics below will present an unbiased platform supporting a guided approach to exploring the benefits of digital transformation, the technical aspects of Industry 4.0 for the manufacturing industry, our entire ecosystem, a methodical design approach to digital transformation, contents & nature of data exchange at various levels, conventional and new system architectures, and various models of system integration to overcome challenges and mitigate any risk faced by industry.

The Journey so Far

Human needs, demands and expectations keep on increasing with the discovery of new technology. Looking back on the history of mankind, previous industrial revolutions have made us improve our quality of lifestyle in many different ways.

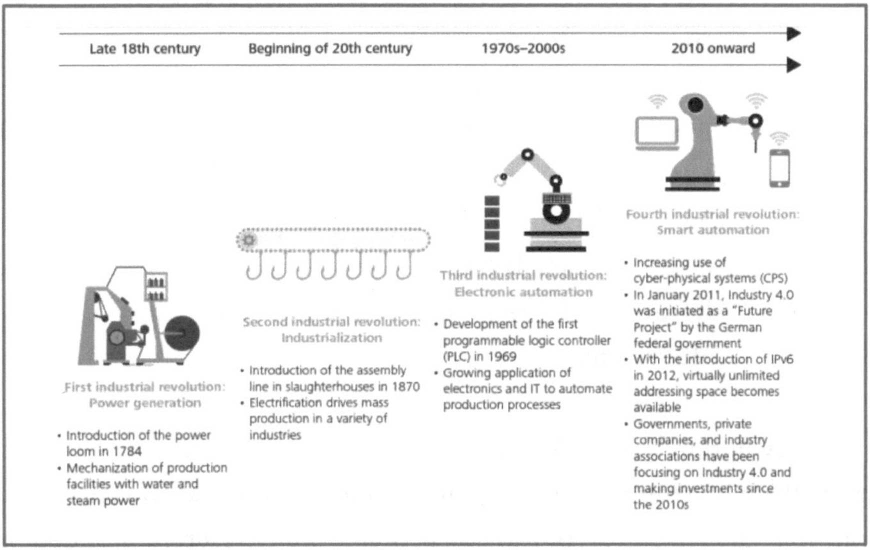

Figure 4: A history of Industrial Revolutions

Each industrial revolution had an immense impact on everything we do and how we live our lives, as briefly stated below:

1. End of 18th Century: Introduction of Mechanical Loom & Steam Engines replaced slow and laborious work practices such as transport by horse-drawn cart, ships rowed by humans, etc.
2. Start of 20th Century: Mass Production based on division of labour powered by electrical energy (Motors, Conveyors), telephone, light bulb, phonograph, and internal combustion engine.
3. Start of 70s: Introduction of Electronics & IT (Electronic Sensors, PC, PLC, SCADA, Robots, Telecom, CT Scanners, X-Ray Machine, Radars, Satellites, 2-D Printers, Laser Scanner and Printers, etc.)
4. Today (Industry 4.0): Introduction of Cyber-Physical Production System (Artificial Intelligence, Machine Learning, 3-D Printing, IIoT, Cloud-based analytical solutions, Blockchain, Connected Machines, Connected People, Connected Homes, Connected Cars, Internet, Global connectivity, ERP, MES, MIS, SCADA, PLC, Robots, Sensors, Wireless Devices, Mobility, etc.)

Each revolution brought its own multi-pronged benefits, opportunities as well as challenges. The first industrial revolution was the Victorian one that kicked it all off by moving from farming to factory production in the 19th Century. The second one ran from around the 1850s to World War I and began with the introduction of steel, culminating in the early electrification of factories and the first spouts of mass production. Closer to home is the third industrial revolution that refers to the change from analogue, mechanical, and electronic technology to digital technology that took place from the late 1950s to the late 1970s.

The definition for Industry 4.0 was first introduced in 2011 at the Hannover Messe trade fair, and was the subject of an Industry 4.0 working group established by the German federal government. MacDougall, William (July 2014) from Germany Trade and Invest (GTAI) defines Industry 4.0 (July 2014) as:

> *"Smart industry or "INDUSTRIE 4.0" refers to the technological evolution from embedded systems to cyber-physical systems. Put simply, INDUSTRIE 4.0 represents the coming fourth industrial revolution on the way to an Internet of Things, Data and Services."*[7]

The fourth, then, is the move towards digitisation and involves three key parts. One is the Internet of Things and cyber-physical systems such as sensors having the ability to collect data that can be used by manufacturers and producers. Secondly, the advancements in big data and powerful analytics means that systems can trawl through huge sets of data and produce insights that can be acted upon quickly. Thirdly, the communications infrastructure backing this up is secure enough to be used by heavy industries.

We have already witnessed increased customer comfort with using innovative technology in the consumer world, such as music & video downloads anytime anywhere, mobile & online banking, ATMs, online shopping, etc. It's the time that we make use of reliable internet for real-time manufacturing operations and manufacturing execution for improved connectivity among an empowered workforce, operations, processes, smart sensors, smart products, and smart devices.

Key Pressure Points of Manufacturing

There are a number of barriers for a business to face when venturing into any existing or new market. It could be attracting new customers – a top pressure point – or cash flow or competitive activity. The following are some of the key pressure points identified:

A. Quality, regulation and compliance
B. Cost control, cash flow and compression, reduce rejection ratio, scraps, and utility management
C. Safety – Automated safety and protection devices, uninterrupted safe operations
D. Innovation – Integrated SCM, Process Improvement, Process Analytics, and efficiency
E. Skilled manpower -finding the skilled manpower, or training existing manpower on a new technology
F. Growth – improvement in expansions, automation and speed and increased productivity
G. Agile, efficient, minimum time to market

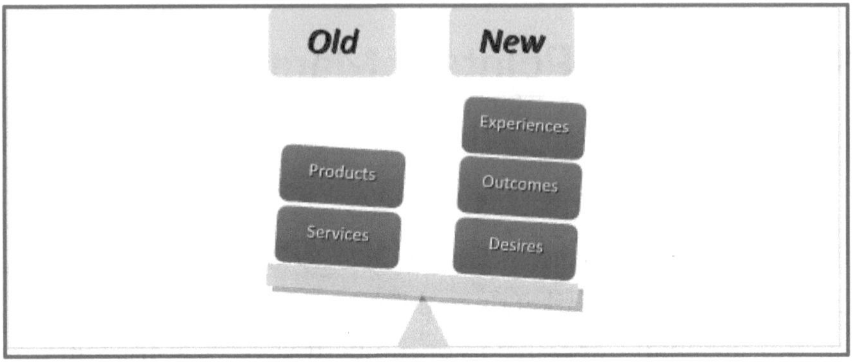

Figure 5: Paradigm Shift in Key Success Factor

H. Sustainability – Predictable & output-driven operations, maintainability and productivity
I. Customer Experiences, Outcomes, Desires
J. Security and Trust of Customers

These factors do vary from market to market, product to product or geographically but present a considerable opportunity to overcome them in a smarter way. On top of this, there are practical situations which can cause a sword of Damocles effect, such as:

- Obsolete & unsafe SCADA, MES, ERP, WMS, SCM Applications
- Several databases with no interconnectivity
- Obsolete reporting and monitoring tools that users find difficult to work with

- No web-based real-time connectivity for remote monitoring and actions
- Current OS may not be compatible with new COTS products available
- Current OS, Applications, DB & IT Network Access are vulnerable to cyber attacks
- Lack of awareness about new technology and business use cases
- Lack of will to change as an organization in terms of process or adopting new technology
- Lack of investment justification (business case)
- Lack of skill set and resources to bring about the change
- Complacency with current state of business affairs

This is where awareness of Industry 4.0 can be helpful in overcoming the barriers and presenting new avenues for growth and sustainability. Anyone who can reduce the above barriers by making smart decisions and adopting intelligent manufacturing properly can come out as a winner!

Key Business Challenges addressed by Intelligent Manufacturing

Pressure points force manufacturers to adopt new technologies and optimize their operations to stay competitive, resilient, and sustainable. There are a number of key business challenges that can be addressed by adopting intelligent manufacturing practices and scalable solutions such as:

- Visualize, monitor, and control operations in real time
- Monitor and control assets, production lines, areas, and plant-level OEE
- Plant Connectivity -diverse assets and equipment & some with legacy systems
- Build common data platform
- Scalable and sustainable – functional and technical architecture
- Flexibility to adapt to multiple business lines and plants
- Convert insights into predictable quality and asset maintenance
- Maintain a high level of security for people, infrastructure, and business
- Optimize energy consumption; control and treat emissions

Primary Business Drivers of Industry 4.0

Whether consumers or businesses, customers are increasingly at the epicentre of the economy, which is all about improving how customers are served. For example, Digital Twin has encouraged many manufacturers to re-think their current business model from a Time-based to an Output-based model.

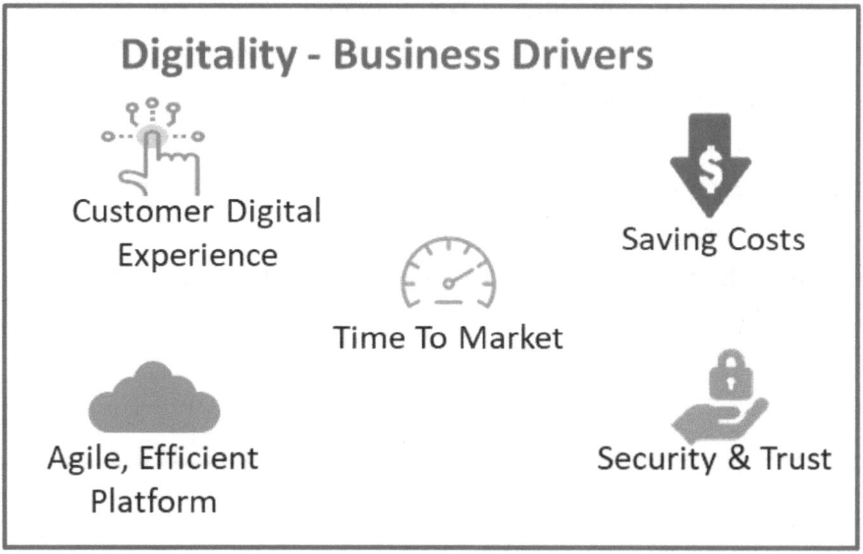

Figure 6: Digitally – Primary Business Drivers of Industry 4.0

Physical products and services, moreover, can now be enhanced with digital capabilities that increase their value. New technologies make assets more durable and resilient, while data and analytics are transforming how they are maintained and consumed. A world of customer experiences, data-based services, and asset performance through analytics, meanwhile, requires new forms of collaboration, particularly given the speed at which innovation and disruption are taking place. And the emergence of global platforms and other new business models, finally, means that talent, culture, and organizational forms will have to be rethought.

As posted by Willamson, J. (2019), *Manufacturing Leaders' Summit: Henrik Von Scheel on moving beyond the hype of Industry 4.0, UK, The Manufacturer [Accessed 21 August 2022]:* "The 17 pillars of 4IR deliver opportunities for growth, innovation, reset competitive landscape, rethink new service models and redefine the next leap in productivity that transforms operational excellence. The Fourth Industrial Revolution is about the fusion of the physical, digital and virtual worlds – Industry 4.0 Infographic".[8]

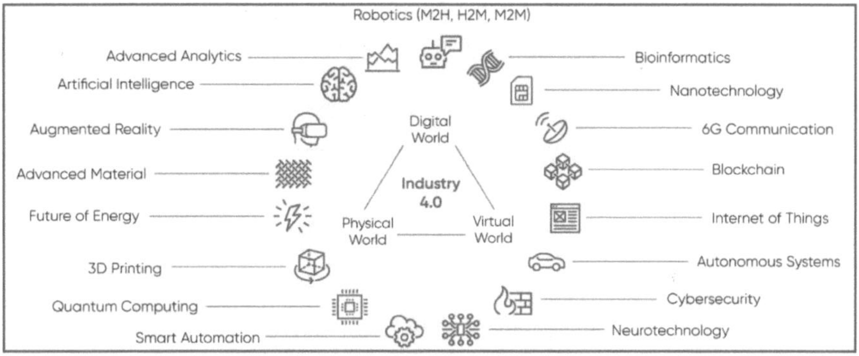

Figure 7: 17 Pillars of Fourth Industrial Revolution 4IR (Source: The Manufacturer)

The newly evolved Ecosystem needed to transform these key drivers into the reality of business as highlighted in the following sections.

Industry 4.0 Ecosystem (Overall)

Of the diagram below, which shows an Industry 4.0 Ecosystem, is self-explanatory. Here, the customer's expectations and satisfaction are at the epicentre and Industry 4.0 is an accelerator or a source of enormous energy that drives the benefits and binds various entities together. All those entities were in action earlier as well, but a new environment is being created by the presence of Industry 4.0 is of enormous value and revolutionary.

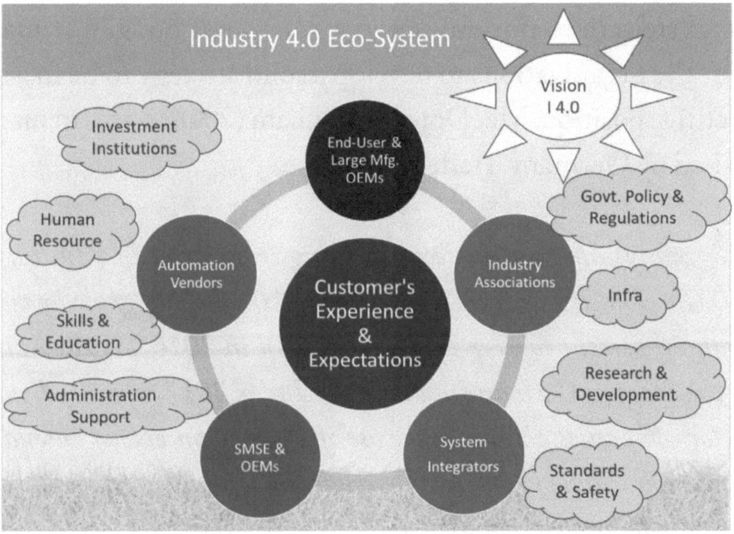

Figure 8: Overall Industry 4.0 Ecosystem

Several of these opportunities didn't exist before and were beyond our imagination. Broad and open ecosystems, including various technologies, telecommunications, skills and the Internet, have enabled a wide range of solutions and unleashed the creativity and innovation of solution-providers worldwide. These ecosystems are shared platforms that use more human and investment capital to create solutions than any single company. The ecosystem is disrupting every aspect of our lives, of our economy, in every industry and country. Different countries, developed or developing economies, are at different level of maturity. Therefore, the transformation to new ecosystem could be the key to manufacturing growth and excellence. It could be the gamechanger as how nations adopt and facilitate this change. And, due to its scale, scope, and complexity, this transformation will be unlike anything humankind has experienced before. Sometimes, an Industry 4.0 ecosystem does not only need a consistent budget. It may also support the organizations to develop a solid strategy to be in a strong competitive position. MacDougall, William (2018) states in his report published by Germany Trade & Invest,

> *"Germany's automotive sector is being transformed by Industrie 4.0. The domestic market for Industrie 4.0 solutions grew by more than 20 percent to over EUR 1.2 billion in 2016/17. According to the Staufen German Industrie 4.0 Index 2017, almost all auto sector companies already make use of "smart" concepts or intend to in the immediate future (compared to 85 percent of respondents in the broader production sector)."*[9]

Recent breakthroughs in technology have allowed us to monitor our health status on the go as well as to get an automated warning to avert any cardiac failure. A number of such Smart edge devices are available in the market today at a very nominal cost. A great example is a Smart Watch, which incorporates processing CPU, sensors, display, and wireless communications supported by a cloud-based application. It may include the wristband, enclosure, 32-bit ARM Cortex-M3, 32-MHz CPU, 128-KB Flash, 16-KB SRAM, 24 Channel x 12-bit analogue-to-digital converter, Bluetooth wireless, three-axis accelerometer, battery, watch and vibration motor. This essentially makes a valuable wireless gadget to help keep our body fit and informed. It is a glimpse into a possible future of automation with edge devices on each machine, such as sensors, actuators, motors, and valves, having embedded processing and communications abilities.

Similarly, on the shop floor, equipment like the Smart Motor, Robot, or Conveyor is critical for the good health of any manufacturing operations and can be monitored directly at cloud level. Artificial Intelligence can help predict the health, remaining life and maintenance needs of smart devices and assets to give advanced warning before failure. Many things that a smart maintenance team used to track using excel sheets can now be automated through a predictive maintenance tool. Further, a prescriptive maintenance application tool can bring wisdom that can advise & alert a local maintenance team to adjust certain parameters of a given piece of equipment in time to provide safe and extended operational life. For example, with an integrated Digital Twin platform, we can see improved operational cost and increase in asset availability.

However, the above-mentioned gadgets and technology can't work alone without a well-planned and engineered overall ecosystem. From cloud / edge-based infrastructure to data security, availability of human resources, skill set, operational intelligence, and high-speed internet connectivity to access the massive amount of data to be exchanged, these are not possible without well-defined strategic goals, policies, infrastructure, planning and the contribution of several entities as suggested in the diagram.

Changing Ecosystem and changing customer expectations enabled by innovation in connectivity solutions embarked on the Digital Transformation journey paves the way to the Digital Enterprise. This evolving market trend is empowered by social, mobile, analytical, and cloud-based technologies. While the overall ecosystem's emphasis is on the required social-economic reforms to bring the desired digital transformation of the entire people-industry-process ecology, we will focus our discussion further on building a Digital Enterprise.

Industry 4.0 Ecosystem for Manufacturing

The Industry 4.0 Ecosystem involves leveraging the latest technologies, digital capabilities, the unprecedented dissemination of IoT (the Internet of Things) and the incredible ability to tap into data anywhere and everywhere in the supply chain.

The level of digital connectivity across the entire end-to-end supply chain, from customers through to manufacturers, distributors, logistics companies, and suppliers of all kinds enables this forward leap to an Industry 4.0 world.

In the attached infographic, we can see that the Industry 4.0 Ecosystem is all-encompassing.[10]

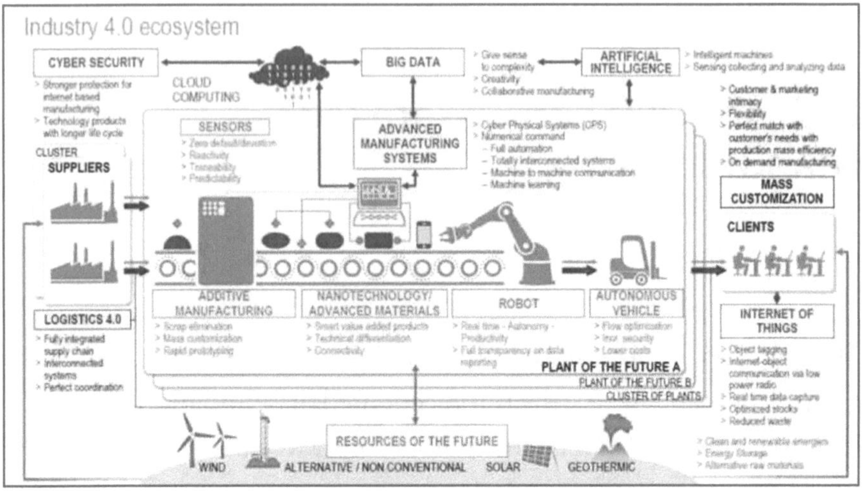

Figure 9: Industry 4.0 Ecosystem

It is critical that companies start to develop their Industry 4.0 Vision, Strategies, and Action Plans. Those companies who can successfully execute and master this level of digital connectivity will have a competitive advantage. This advantage will manifest itself in terms of advanced capabilities for their customers, improved operating performance and productivity, better financial results, and the platform for future, unprecedented growth.

A key trend is the development of technology-enabled platforms that combine both demand and supply to disrupt existing industry structures, such as those we see within the "sharing" or "on-demand" economy. These technology platforms, rendered easy to use by the smartphone, convene people, assets, and data—thus creating entirely new ways of consuming goods and services in the process. Produced goods with embedded intelligence can be tracked & monitored for enhanced user experience and proactive maintenance.

While covering the technicalities and benefits of an entire digital supply chain is a very elaborate topic to cover in this book, it would be prudent to touch primarily on shop-floor layers in a manufacturing environment and associated digital transformation as part of the ongoing trends under the Industry 4.0 Ecosystem.

Revolutionary in nature, Industry 4.0 (IIoT)—Introduction of Cyber-Physical Production System – brings many innovative technologies combined together that are so impactful and ready to use, as suggested below:

- Big Data, Artificial Intelligence, Machine Learning
- 3-D Printing
- Edge/Cloud-based computing
- Digital Twin
- Augmented Reality (AR) / Virtual Reality (VR)
- Connected Machines
- Connected Products
- Empowered or Connected Workforce
- High Speed Internet Connectivity
- Asset performance Management (APM) using Digital Technology
- Enterprise Resource Planning (ERP)
- Product Lifecycle Management (PLM)

- Warehouse Management Solutions (WMS)
- Manufacturing Execution Systems (MES)
- Supervisory Control and Data Acquisition (SCADA) System
- Programmable Logic Controllers or Machine Controllers
- Connected Robots, Cobots or AGVs
- Smart Sensors, Wireless Devices, Mobility Solutions, and so on.

With the everyday inventions in digital technology – gadgets, equipment, materials, processes, innovations, improved infrastructure, Mobility Solutions – manufacturing competency has become a crucial norm of global leadership. Covering all topics in detail would not be a frugal exercise as exponential changes in technology, and the co-existence of multi-dimensional and multi-domain solutions will be overwhelming to cover in one book. Each vendor may have its own solution or product to suit a specific industry or domain, lack of global standards in an evolving landscape, and so on. However, it would be worth mentioning the high-level solutions, the challenges to address, tools to use, methodology to follow, and benefits that the manufacturing industry can gain.

Industry 4.0 and Intelligent Manufacturing Initiatives

Countries across the globe consider the manufacturing sector to be one of the key sectors to increase the economy of the country. IoT is reshaping the manufacturing as well and commercial space. With the IIoT, the manufacturing industry is expected to achieve a number of benefits in terms of optimizing their production process through reduced lead time, increased efficiency and reduction in time taken for the product to reach the market. The following use cases involving Industry 4.0 have presented an enormous opportunity to gain for the industry:

- Condition Monitoring of Equipment
- Predictive Maintenance
- Production Automation – Productivity Suite
- Intelligent Supply Chain – Supply Chain Optimization
- Energy Management – Reduced Carbon Footprint
- Additive Manufacturing
- Pick-to-Light System
- Paperless Reporting
- Digital Twin – Process/System/Asset

- Augmented Services (AR/VR) – for Maintenance and Training
- Remote Service

In North America and Europe, there has been an increased interest in achieving the Industry 4.0 and Intelligent Manufacturing in the near future. IIoT is seen as one of the technologies for enabling the same. As per a survey and subsequent white paper published by Orange (Dec 2018), "the most popular use cases of companies already implementing IIoT solutions, are those related to supervision & maintenance. The following figure gives a non-exhaustive list of identified addressed use cases."[11]

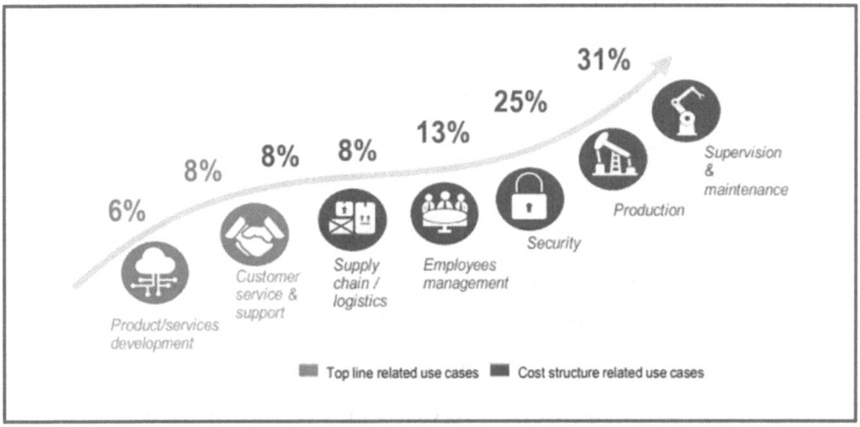

Figure 10: IIoT Use cases (Source: Orange_5G_Iiot_whitepaper_dec2018)

IoT in Intelligent Manufacturing will provide a gateway for commercial ICT vendors to enter the manufacturing space as it is considered to be one of the key enablers for the realization of Intelligent Manufacturing.

Factories of the future are concepts leading to improved productivity and efficiency. With the advent of IoT in Intelligent Manufacturing, existing industrial suppliers are set to witness increased competition from commercial IT vendors in the future.

In another dimension, industrial machines and the goods they turn out are increasingly packed with internet-connected sensors. Manufacturers are thus able to gather data on how their machines perform out in the world. Their intimacy with the product and the amount of data they accumulate gives them a base from which to sell services which no third party can match. A maker of cars, or wind turbines, or earth movers can use data from every product it has made to work out what is going on with any one of them, and thus increase the value to the user—who is increasingly likely to pay for the service that the manufactured object offers, rather than the object itself. The car industry, for most of the 20th century the archetype of metal bashing, increasingly sees its future in the provision of "mobility services" rather than as a seller of boxes with wheels at the corners. Running their own fleets of cars with which to offer autonomous or shared rides looks to many like the wave of the future—and possibly a very profitable one.

The enthusiasm for moving into services extends well beyond the makers of high-end machinery with whom the trend started. Adam, Henrik at Tata Steel in Europe says

> *"he has a team of experts able to intervene in a customer's production line and improve their manufacturing performance and yield by*

> *specifying the best type of steel to match processing capability and market ambitions."*

Similarly, Lafarge-Holcim, a cement-maker, says

> *"its product can be delivered as a service. Increasingly complicated cement structures require experts to advise on design, use of specialist products and the logistics of pouring a continual stream of the stuff."*

Skills and Education

No doubt that Automation will decimate cheap labour job market. The answer is to enhance skills through education reforms.

The warning bells sounded very early highlighted the threat from automation to lesser skilled occupations in manufacturing and services. It's not just about displacement of existing occupations. It puts to test the very developmental model that Asian or African countries used to claw back into the core of the global economy, from the peripheral position imposed on them through a combination of technological change and colonisation.

The model rested on state-supported industrialisation and a focus on export orientation using lower production costs as a competitive edge. Lower production costs were largely on account of lower labour costs in the early stages, and later increases in labour productivity. A productive and highly automated manufacturing facility may no longer need cheap labour to do the shop-floor menial work anymore.

But with increasing automation, the terms of trade would again shift in favour of owners of intellectual capital and technology and diminish the returns to labour. This would be a great disadvantage to any country not just development countries with large, young working populations. The key to create a healthy job market without losing competitive edge is to re-skill the workforce, upgrade the education system with new technology, new syllabus and learning practices. Remember the transition from conventional Typewriter to PC/Tablet with word editor was neither so simple nor happened overnight.

Capturing a larger share of the Industrial Revolution 4.0 would require investment in high-quality skills related to applied science and technology, research, engineering, quantitative and social analysis, design and product development. Since shopfloor activities that would still be done by humans would require high familiarity with technology and analytical abilities, workers would need to have educational levels currently available to college graduates and advanced industrial training institutes. The next generation industrial policy must be intrinsically linked to reformed education policy.

Big Data for Manufacturing

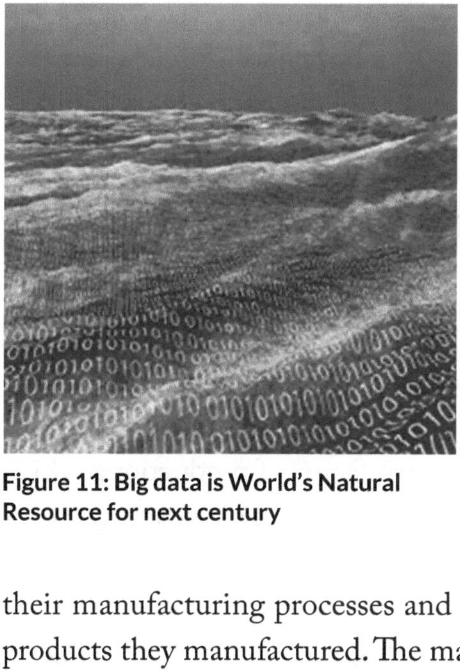

Figure 11: Big data is World's Natural Resource for next century

The manufacturing sector was an early and intensive user of data to drive quality, efficiency, adopting information technology, automation to design, build, and distribute products since the dawn of Industrial Revolution 3.0. In the 1990s, manufacturing companies racked up impressive annual productivity gains because of both operational improvements that increased the efficiency of their manufacturing processes and improvements in the quality of products they manufactured. The manufacturing sector has been the backbone of many developed economies. Increasingly, global and fragmented manufacturing value chains create new challenges that manufacturers must overcome to sustain productivity growth. In many cases, technological change and globalization have allowed countries to specialize in specific stages of the production process. To continue achieving high levels of productivity growth, manufacturers will need to leverage large datasets to drive efficiency across the extended enterprise and to design and market higher-quality products. Tymchuk, Igor (Oct 2020) of software development service firm "N-iX" notes in his article "Big Data in manufacturing made simple: Trends, tips, success stories"[12],

> *"Big Data is changing the face of manufacturing as we know it. Gone are the days when data scientists need to mine, organize, and manage volumes of diverse data manually or run experiments to test the hypothesis. With Big Data, manufacturers can predict business outcomes, enhance the quality of their products, reduce equipment downtime, and optimize costs."*

Traditional economics, based on an industrial economy, teaches that there are three factors of production: land, labour, and capital. The December 27, 2012, issue of the *Financial Times* included an article entitled "Why 'Big Data' is the fourth factor of production," which examines the role of big data in decision-making. According to the article:

> *"As the prevalence of Big Data grows, executives are becoming increasingly wedded to numerical insight. But the beauty of Big Data is that it allows both intuitive and analytical thinkers to excel. More entrepreneurially minded, creative leaders can find unexpected patterns among disparate data sources (which might appeal to their intuitive nature) and ultimately use the information to alter the course of the business."*[13]

Big Data has become the new power for industry and businesses in the 21st century. There are three main attributes for Big Data identified by Russom, Philip (2011), TWDI Research in his The Three Vs of Big Data Analytics.

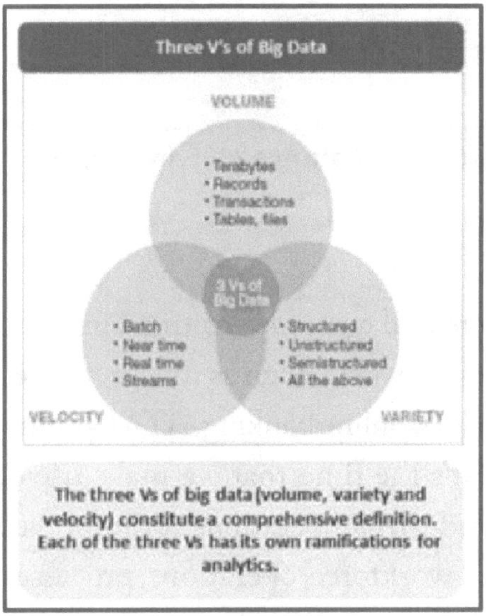

Figure 12: Three Vs of Big Data

[14]Big data's first main attribute is the volume. Some quantified data by counting their records, transactions, tables or files, but some found it more useful to quantify big data in terms of time. For example, in the U.S., some prefer to keep data available for legal analysis for seven years which is the statute of limitations. The second attribute is the variety of data. This happens because data come from a variety of sources like logs, streams, social media, text data, and semi-structured data from B2B processes. The last attribute of big data is the velocity which refers to the low-latency, real-time speed at which analytics need to be applied. **In Figure 12, heading, change 'Three V's' to 'Three Vs'.*

Later, we will discuss more use cases of Big Data in several aspects of the manufacturing industry.

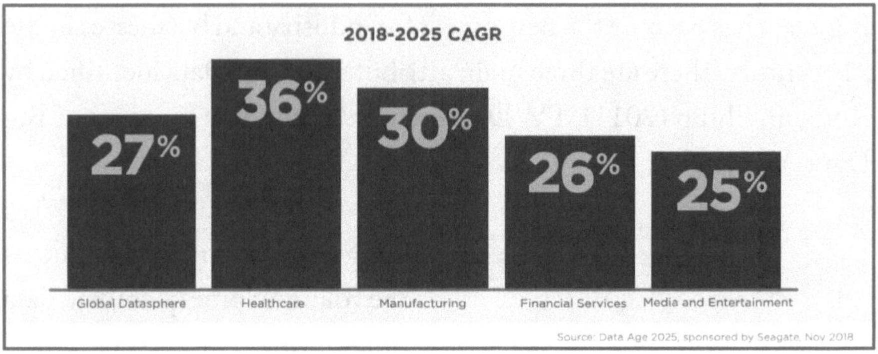

Figure 13: Big Data Market CAGR figures (Source: Data Age 2025, by Seagate, Nov 2018)

We have already witnessed the increased customer comfort in using innovative technology in the consumer world, such as music & video downloads anytime anywhere, mobile & online banking, ATMs, online shopping, radio taxi services, etc. It's the time that we make use of reliable technology for real-time industrial operations for improved connectivity among an empowered workforce, operations, processes, smart sensors, smart products, and smart devices. This is all possible because of data exchange over the cloud among various entities in a pre-defined framework.

Water Cycle Analogy of Data Cycle

Within Industry 4.0, data movement may be compared to a water cycle but not literally as described below. Many terms of the water cycle model, including lakes, oceans, clouds, pipeline or reservoirs, are frequently used by industry to draw a parallel relationship and functionality and to represent closely matching meaningful relationships.

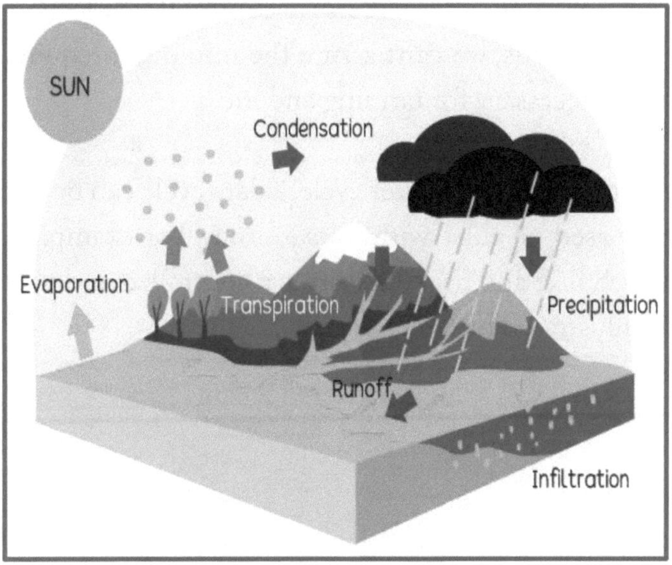

Figure 14: Hydrosphere and water cycle

Think of all the water in the oceans and seas, including all the frozen water and ice (cryosphere). Also, think of all the lakes, lagoons, rivers, and ponds, as well as water in the water table beneath the surface of the earth. They are all part of the hydrosphere and together they cover more than 70% of the surface of the earth.

The hydrosphere is also in infinite processes every day with the Sun and Earth as driving factors. The water cycle is one way to understand how the hydrosphere functions and supports other spheres. However, it's a natural process and beyond human control. Sometimes, heavy rains and cloudburst can cause havoc for human beings. On the other hand, clouds can miss the timeline or direction during monsoon resulting in droughts in several regions. Unless we control the monsoon with the movement of clouds, we can't assure the minimum required rainfall over a region necessary for farming and life.

In contrast to the natural water cycle, a data cycle can be defined and controlled based on rules within its journey. For example, data lake to data warehouse and back to users is a controlled and well-defined journey as illustrated below.

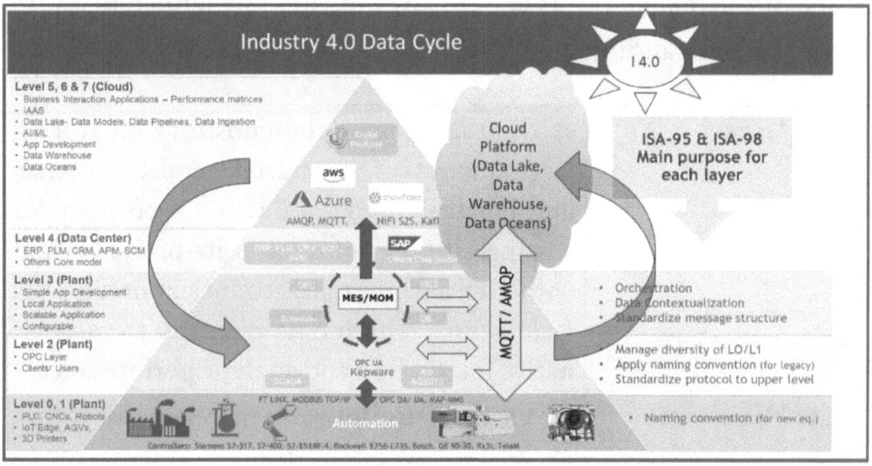

Figure 15: Manufacturing Data Cycle

As per industry expert Nicholas Samuel (Data Lake Architecture: 10 Critical Aspects, October 21st, 2021, accessed on 11-09-2022): "Data Lake Architecture is the product of planning and designing scalable storage to handle a growing amount of data while providing faster insights"[15].

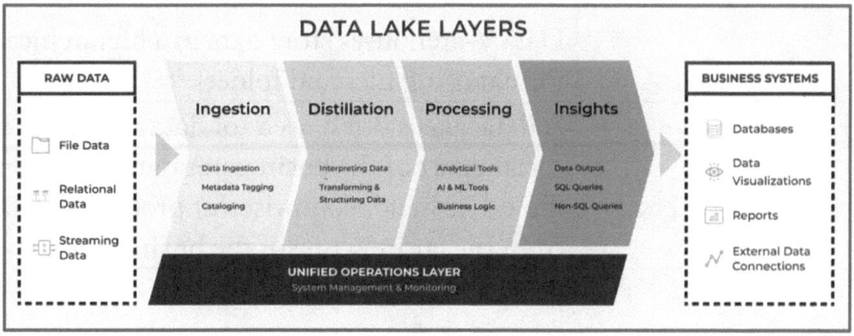

Figure 16: Data Lake Layers and Pipeline

Some of the related terms adopted by industry are defined below for a quick understanding:

Data Lake	A Data Lake is a data repository for storing large amounts of Structured, Semi-Structured, and Unstructured data. It is a repository for storing all types of data in its native format without fixed limits on account size or file. Data Lake stores a high data quantity to increase native integration and analytic performance.[16]
Data Warehouse	A Data Warehouse is a repository that exclusively keeps pre-processed data from a Data Lake or many databases. ETL (Extract, Transform, and Load) operations are used to arrange data in multi-dimensional structures so that Analytics workflows using Data Warehouses can be accelerated. Business Intelligence specialists and Data Analysts can generate reports and develop dashboards using the data housed in a Data Warehouse. Data Warehouses store data in a hierarchical format using files and folders.[17]
Data Ocean	A data lake is often used for data specific to a certain part of the business. A data ocean, on the other hand, comprises unprocessed data from the entire scope of the business.[18]

Data Reservoir	Some businesses use the concept of a data reservoir to distinguish between a totally unrefined data lake and a data reservoir that's been partly filtered, secured, and made ready for consumption. Imagine a lake that's been drained and filtered into a reservoir, creating a source of potable drinking water.[19]
Data Mart (similar to Energy Drink, Soft Drink, Potable Bottled Water, Tap Water)	A data mart is a subset of the data warehouse tailored to the needs of a specific team or line of business. Think of it as a storage room within your warehouse used to store only data within a specific scope.[20]

Table 1: Common Terms and Definitions for Data operations

Each Data movement can be controlled, monitored, filtered or aggregated, analysed and presented to deliver a certain value to the consumer as needed in a Data Cycle.

Fabric of Industry 4.0

Industry 4.0 is evolved as a Cyber-Physical system that enables intelligent communication of many entities over the internet from corporate level to device level or from supply chain to end-customer involving following process domains:

- Customer Chain – Customer Excellence
- Product / Service Lifecycle – Innovation Excellence
- Supply Chain Excellence
- Manufacturing Excellence

Conventional Enterprises have information available in Silos and exchange of information is often a slow and painfully complex process as illustrated below. This often results in an elongated decision-making process and longer time to bring product/ changes to market or respond to quality defects.

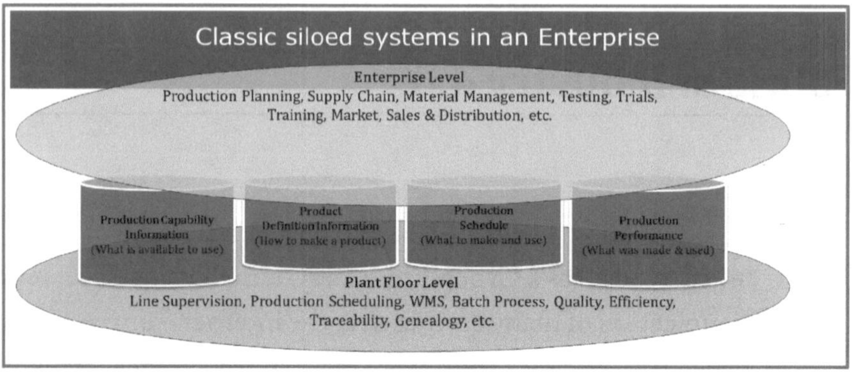

Figure 17: Classical Siloed System in an Enterprise

It requires a huge communication infrastructure and computing power reliable enough for the industrial applications consisting of large number of smaller inter-connected networks to collect and distribute meaningful data. While currently, operations may be optimized within various system silos and organizational boundaries, advance enterprises will be well-connected, data-driven and integrated across various System Silos and domains in a near-instant scenario to offer greater values.

Connectivity plays a crucial role and impacts which process domain within the enterprise will excel. Connectivity is at the heart of the Industry 4.0 reference model as it enables the information exchange across various process domains in manufacturing.

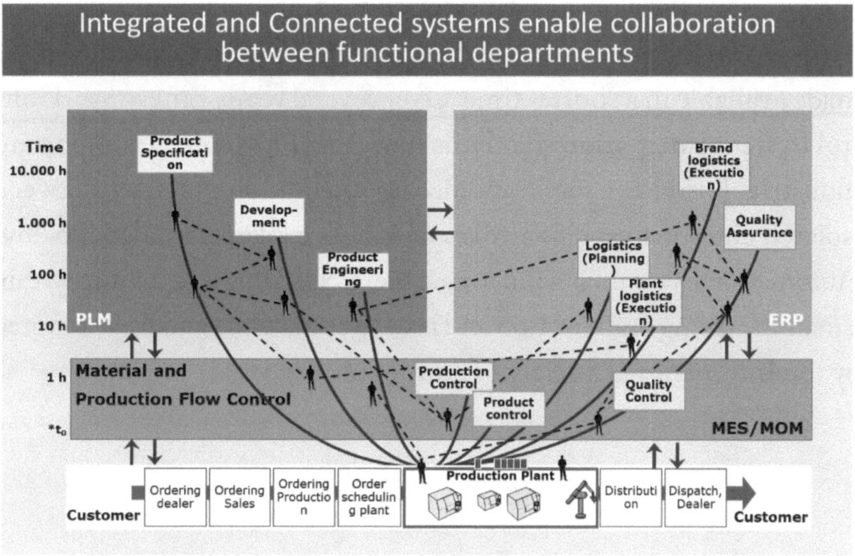

Figure 18: Integrated and Connected Systems in manufacturing process

The following entities are being connected through the implementation of Design Principles and Technologies:

- Connected Enterprises
- Connected People
- Connected Things
- Connected Applications

Connectivity within the enterprise and with the outside world is key to any connectivity model. On one side, it enables customers to provide direct feedback to the product designers and thus customers influence product design. On the other side, Smart products provide online customer/product profile data thus Smart products generate usage data that enables individually customised products or services to be made available in a shorter time. Over several years, Enterprise-Control System Integration standards have sought to solve an important industrial business issue: normalizing integration practices between isolated enterprise and control systems. As explained in detail below, 'Automation Pyramid' is a functional hierarchy defining activities within a manufacturing organization and is well-established and well-adopted by the International Society of Automation (ISA).[21]

Five Layers of Automation Pyramid

Until recently, the Automation Pyramid was a classical pictorial representation of the different levels of automation in a factory. Conventionally, it was a great way to make sense of the entire complexity in a factory. However, with the introduction of Industry 4.0, the traditional hierarchical manufacturing structure will appear to be networked differently.

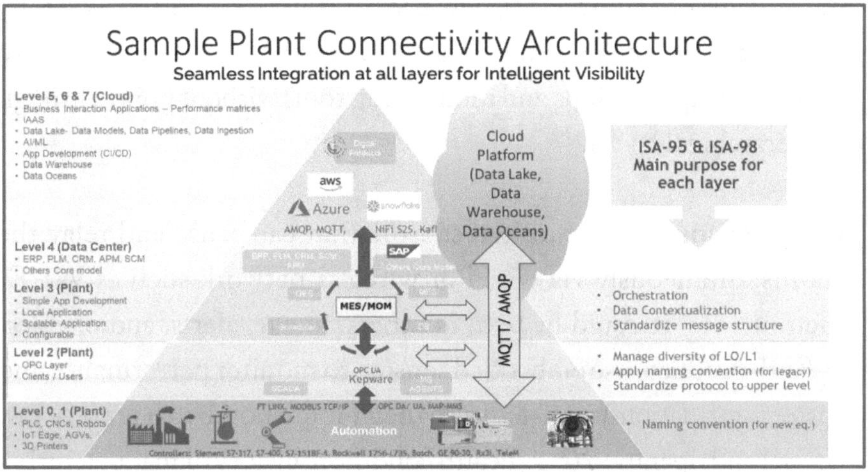

Figure 19: Automation Pyramid and its Functional layers

The various layers of a conventional Automation Pyramid are explained below for quick understanding.

Level 0 – Field Level/Production Process

This level has devices, actuators, and sensors that are found in the field or on the production floor. The field level is the production floor where physical work and monitoring happens. Electric motors, hydraulic and pneumatic actuators to move machinery, proximity switches to detect movement of certain materials, and photoelectric switches that detect similar things are some examples of field-level devices. With the introduction of Industry 4.0, new use cases have emerged to directly monitor Smart Sensors and Smart Devices by connecting them to cloud-based applications and monitoring tools without going through the hierarchical levels.

Smart Sensors — Connected sensors that can sense and relay the data instantaneously via wired or wireless network such as Wi-Fi, Bluetooth etc. It could be text, numbers, image, alerts, audio, video, etc. End-products use embedded sensors to monitor performance and register information associated with their production, batch process, raw material and operation (e.g., health condition, history of manufacturing and service life). This gives them a unique identity. With revolutionary nano-technology used in tiny sensors, the world is witnessing an increased use of signals when transmitted to a remote location over the internet in real time either for immediate action, statistical analysis or recording purposes.

Smart Devices — Connected Video Cameras, 3-D Printers, Autonomous Vehicles, Drones, Robots, Controllers, Variable Speed Drives (VFDs), Smart Motors, Bar-code scanners, Video-Cameras, Smart Watches, Smart Transformers, Smart Phones, Smart Valves, Smart Pressure Transmitters, and Smart Level Transmitters have become the norm in business.

The invention of these smart connected devices has revolutionized the entire business community – News, Media, Commercial, Consumer, Industrial, Medicine, Agriculture, Infrastructure, Horticulture, Transportation, Telecommunication, Biotechnology, Legal, Transportation, Security, Safety and Social worlds. These devices, combined with other components of Industry 4.0, could become smart enough to talk to each other, reconfigure themselves, transmit the data on demand and keep learning based on the statistical analysis.

Level 1— Control Level/Sensing and Manipulation

Machine or process controller devices such as PLCs , DCS, CNCs, AGVs, or PID Controllers operate at this level with nano or milli-sec response timings.

At this level, you control and manipulate the devices in the field level that actually do the physical work. They take in information from all of the sensors, switches, and other input devices in raw form to make decisions in micro-seconds on what outputs to turn on to complete the programmed task. PLCs contain a processor, memory to hold their programming and other data and input and output modules. Simultaneously, these machine controllers or edge devices transmit the raw data to the upper layers for further monitoring and processing. The key to sending raw data is a proper naming convention, low latency, and, in some cases, time-stamping to identify and relate data to other entities –Batch ID, Process IDs, Product Types, Product ID, Sub-Areas, Line, Shift Name, Operators, Team, and so on.

A common example is a cruise controller. A cruise controller can be set to whatever speed you want, then a set of sensors and the computer in the car will tell it when to accelerate or decelerate according to the set speed. However, the same data can be sent to the cloud while it can be linked to the car's VIN number, driver name, car status, fuel consumption, local regulations, fuel efficiency, location, weather conditions and so on.

A common industrial PID-controlled item is a heater. Many systems in manufacturing plants have to be heated. We control this with a PID block within the PLC. When a set point is entered, the PID will determine when a controller needs to turn the heater on and off to maintain a constant temperature.

Level 2 – Supervisory Level/Monitoring and Supervising (SCADA)

This level is known as the supervisory level. Where the previous level utilizes PLCs, this level utilizes SCADA or Cloud-based apps or other data collectors. SCADA is the acronym for a supervisory control and data acquisition system.

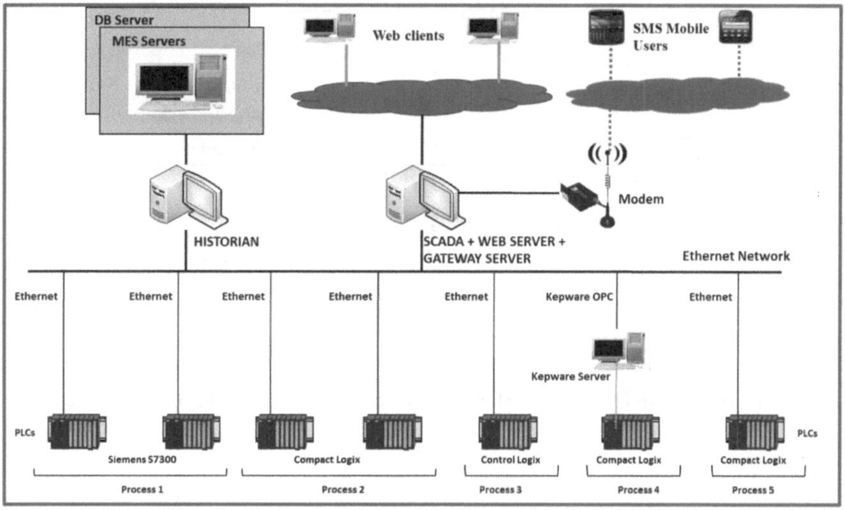

Figure 20: SCADA system at level-2

SCADA is essentially the combination of the previous levels used to access data (also acts as data collector) and control systems from a single location. It usually adds a graphical user interface, called an HMI, to control functions remotely.

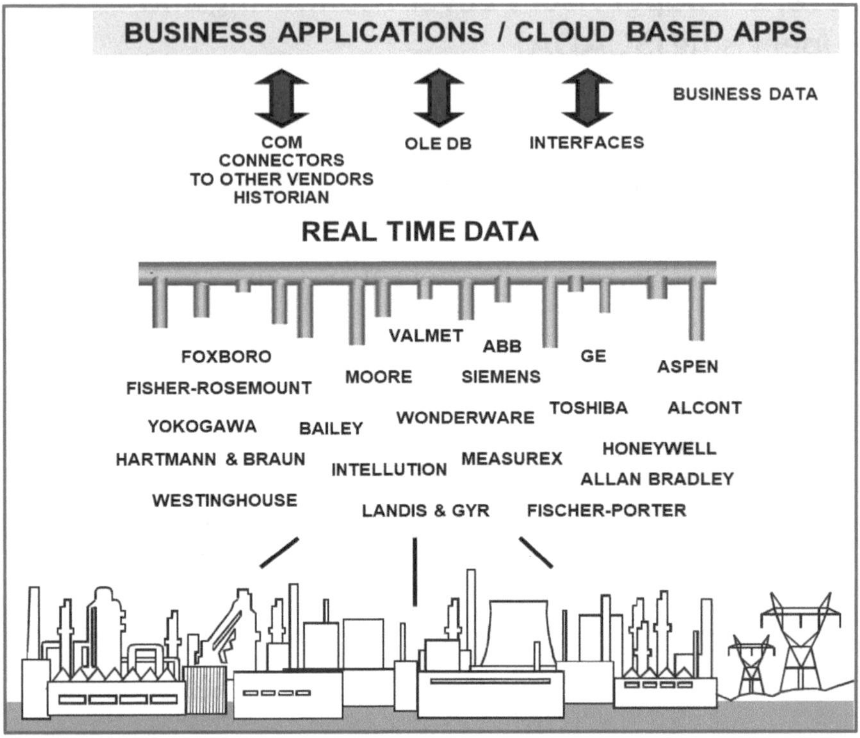

Figure 21: Real Time Data from SCADA /MES sent to upper-level Business Applications

The important thing to remember about SCADA is that it can monitor and control multiple systems from a single location. Unlike the conventional manner, it isn't limited to a single machine or piece of equipment but can be used to monitor a complete line, area or plant or item of equipment (static or mobile) scattered over multiple geographical locations using the Internet or other remote communication methods to collect data.

Level 3 — Execution and Operations Management

The fourth level of the automation pyramid is called the planning level – Execution and Operations Management. This level utilizes a computer management system known as MES or manufacturing execution system. MES monitors the entire manufacturing process in a plant or factory from the raw materials to the finished product.

A manufacturing execution system (MES) provides the foundation for operations to schedule, manage, and track each step during production. This allows management to see exactly what is happening and allows them to make decisions based on that information to efficiently operate the facility. They can adjust raw material orders or shipment plans based on real data received from the systems. This is the layer for Data Orchestration and Contextualization. All the raw data available from the shop floor is aggregated in a relational database for meaningful reporting and presentation to the top layers' KPIs to make effective decisions.

Modern MES systems are cloud-based and available as a 'Software as a Service (SaaS)' solution, using software that runs over the Internet in a public cloud. Using cloud MES solutions, all involved personnel can access real-time operational data at any time. This connects departments and helps businesses monitor manufacturing operations, and supply chains to form integrated strategies based on the data they've collected. Synchronization between MES and ERP (Enterprise Resource Planning application) can be straightforward, and it provides the ability for an untethered workforce.

Due to its critical and complex nature, more insight into MES system functionality will be covered later in the book.

Level 4 – Enterprise Level (Business Planning and Logistics)

The top of the ISA-95 automation pyramid is what is called the Enterprise level. This level uses the companies' integrated management system such as Enterprise Resource Planning (ERP), Product Lifecycle Management (PLM), or Supply Chain Management (SCM).

This is where a company's top management can see and control their entire operations. ERP is usually a suite of different computer applications that can see everything going on inside a company. It utilizes all the previous levels' technology plus some more software to accomplish this level of integration. It needs careful planning, selection of software, configuration, and architecture to meet its strategic goals at Enterprise level.

As is well defined in an article by Gills, Alexander S. in TechTarget,

> *"An enterprise architecture (EA) is a conceptual blueprint that defines the structure and operation of organizations. The intent of enterprise architecture is to determine how an organization can effectively achieve its current and future objectives. Enterprise architecture involves the practice of analysing, planning, designing and eventual implementing of analysis on an enterprise."*[22]

This allows the business to be able to monitor all levels of the business from resource planning, production planning to forecast, design to manufacturing, procurement, sales, marketing, finance and payroll, plus many others. The integration of the ERP promotes efficiency and transparency within a company by keeping everyone informed with a single truth.

Level 5, 6 & 7 – Cloud layer

Cloud is what forms the digital transformation part of the factory where data from any of the previous levels can directly be fed into a Data Lake or other applications and insights can be gathered. Here is where the next level of efficiency and operational excellence is achieved as illustrated in the diagram. This is a newly evolved layer and is beyond traditional layers of the ISA-95 Automation Pyramid. Also, it influences the way conventional layers were networked earlier in hierarchy.

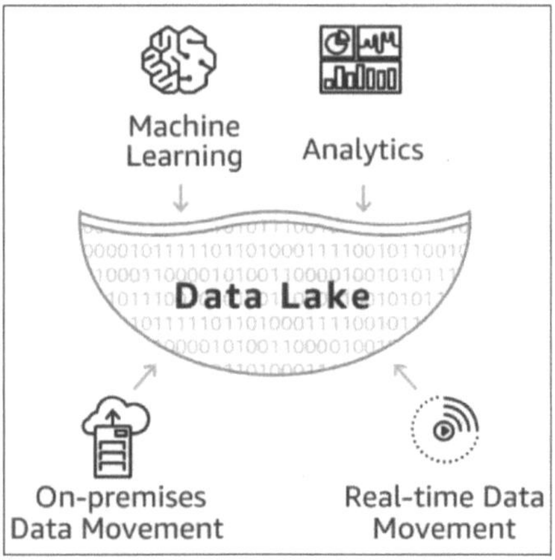

Figure 22: Data Lake operational excellence

Within a short time, Cloud Native (new term used by industry) has become a driving trend in the industry. It's a new way to construct large, complex systems. The approach takes full advantage of modern software development practices, technologies, and cloud infrastructure. Cloud native changes the way you design, implement, deploy, and operationalize systems.

In an article posted on webpage of OPCFoundation.org by SAP SE notes,

> *"The market increasingly demands individualized products. Product life cycles are getting shorter and average lot sizes are getting smaller. Considering this, cost efficient production in a globalized and resource limited environment requires a highly flexible (IT) infrastructure that also works well in cross-enterprise networks."*[23]

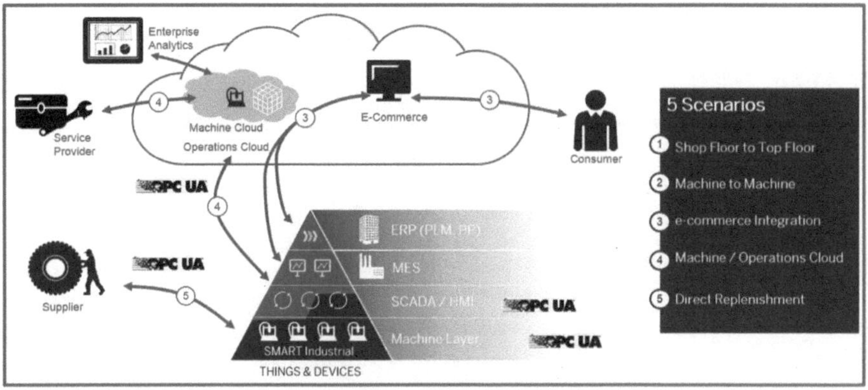

Figure 23: Reshape the Automation Pyramid (Source: OPCFoundation.org **webpage by SAP SE)**[24]

More details on design consideration for scalable applications using Cloud Native technologies will be covered in the following sections.

Design Consideration for Cyber-Physical System

A 3-D map explains the design approach of Industry 4.0 in a structured manner to deliver better performance to manufacturing, business challenges and enhanced customer experiences. Recently, the RAMI 4.0, Reference Architecture Model Industry 4.0, was developed by the German Electrical and Electronic Manufacturers' Association (ZVEI) to support Industry 4.0 initiatives, which are gaining broad acceptance throughout the world.[25]

Instead of a 3-D map, it could have been 7-D for more practical implementation reasons as mentioned in the Hierarchy section. RAMI 4.0 is a three-dimensional map showing how to approach the issue of an Industry 4.0 in a systematic and structured way. RAMI 4.0 is a unified model for all the components that ensures all the participants involved in an Industry 4.0 ecosystem share data & information efficiently and effectively. RAMI 4.0 maps all the players of the connected industry to three axes of definition as listed below:

- Life Cycle value stream

- Hierarchy levels (Automation Pyramid)
- Architecture Layers

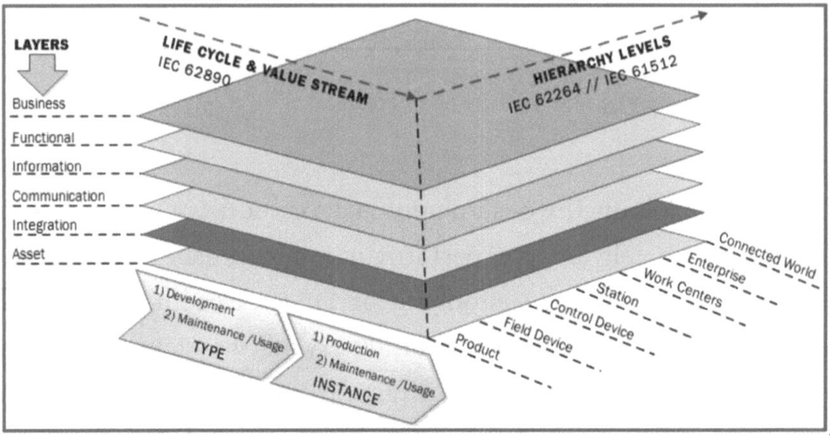

26

Figure 24: The RAMI 4.0, Reference Architecture Model Industry 4.0 (Source: www.zvei.org)

"Connected World" level is an additional level introduced in the hierarchy structure as compared to the conventional ISA-95 Automation Pyramid. Plus, all the data from these hierarchy levels can be simultaneously consumed by various layers such as Business, Functional, Asset, Communication, Information, and Integration; thus, avoiding siloed exchanges of data. With the third dimension added, it ensures Lifecycle and Value Stream of Production/Product Development and Maintenance/Usage domains as elaborated in the below sections.

Lifecycle Value Stream

The lifecycle value stream axis is divided into Type and Instance. The Type is divided into Development and Maintenance/Usage, while the Instance is divided into Production and Maintenance/Usage. A type represents the initial idea of the product development, while each manufactured product represents an instance of that type. The value stream in the totally digitized production can be viewed in conjunction with value-adding processes, since it enables linking of purchasing, production planning, logistics, quality, customers, and suppliers. In simple words, if a product is in a developmental state, then we refer to it as a "Type". Once it moves into production, then it becomes an "Instance". Whenever a product is redesigned or a new feature is being added to it, then again, its state turns to "Type".

Hierarchy Levels

The Industry 4.0 architecture at hierarchical level shows a functional assignment of components as explained above in the section on Automation Pyramid. This axis within an enterprise or factory follows the IEC 62264 and IEC 61512 standards.[27]

The level over and below the IEC standards area represents steps further and also describes groups of factories, collaboration within external engineering firms, component suppliers and customers. Therefore, the hierarchy levels are: product, field device, control device, station, work centre, enterprise, and connected world as also defined in the above section on the Automation Pyramid.

Architecture Layer

The architecture layer includes sub-layers as mentioned above. This layer enables the development of Industry 4.0 software solutions in a consistent way so that different and mutually dependent manufacturing operations are interconnected considering the physical and the digital world.

RAMI 4.0 breaks down complex processes into packages, making it easily understandable, and includes, by design, data privacy & IT security. It addresses and answers all the problematics about semantics, identification, functions, communication standards, internationalization and partnering for the smart factory.

With RAMI 4.0 architecture, the factory is not an overlay of layers but a network of interaction between smart products and the connected world. Functions are distributed to all participants allowing more flexible operations between systems and machines. To connect physical things to the Industry 4.0 world, the RAMI 4.0 provides an administration shell which comes to be placed over the system; the system being a sensor, a machine, a unit or a plant. Also, with every asset having its own administration shell, several assets can then form a unit with a common administration shell for higher-level communication and management. Once connected to the physical system, the shell serves as the network's standardized communication interface to share all data and information about the asset.

As described, the function of RAMI 4.0 is to transform physical objects into their interoperable digital twin. Bringing the industrial world into the IoT, RAMI 4.0 is the root of the Industry 4.0 revolution.

Example: Edge analytics solution architecture in the context of the RAMI 4.0 showing architecture layers.

Hierarchy Levels: The edge analytics solution architecture in the frame of RAMI 4.0 is applicable at component, machine or production process level. In this sense, it can be implemented in flexible smart systems and machines capable of interacting and communicating across the hierarchy levels through a network. The implementation of the architecture in a "Connected World" (i.e., connected factories with integrated edge analytics processes) would require its use by all of them in order to create synergies (e.g., between a factory and its supplier of raw materials or spare parts for maintenance).

There could be three hierarchy levels instead of one, namely – Physical, Organisational and Logical Hierarchy. It makes more sense to define relationship among nodes of three hierarchies. For example, a physical workstation can be related to an operator and an assembly line or sub-area can be related to a team. Similarly, a logical PLC unit can be related to a station(s) or an assembly cell. It will provide insight into manufacturing operations by mapping various KPIs (discussed later in the book) to their respective logical, physical or organisational entity or node.

Industry 4.0 technologies are developed to enable smart prediction systems to flexibly make decisions during the production process. In Industry 4.0, most of the operational decisions are made according to the experience of supervisors. However, how to analyze the collected information and provide the insights and meaning to help the users to make better decisions is a crucial issue for the current prediction systems. Though the data collection process is simple and easy to implement by using sensors and data acquisition software, without the data analysis, collected data is useless. It is necessary to calculate, store and analyze the data to make it meaningful and useful for shop-floor operations. Using the big data analysis and cloud computing platforms, some of the important insights or upcoming events in the machines can be discovered. The development of the digital system and decision support system will transform the information into visualization which users can see and understand.

The characteristics of Industry 4.0 ecosystem in the 21st century can be applied to a smart prediction system as described in the table below:

Characteristics of enabling technologies	Description
Internet of Things (IoT)	a) Ability to connect and manage machines / equipment b) Sensor data collection from machines in real time
Big data analytics	a) Allow real-time use of analytics tools for large amounts of data b) Real-time decision-making support
Edge computing / Cloud computing	a) A model for managing, storing, and processing prediction system data on edge or in cloud b) Real-time commands to machines based on data analysis results computed on edge or in cloud
System integration	a) Allow for different computer or machines to be linked together b) Enabling actual communication and passing on of IoT data from machines to IT systems

Table 2: Characteristics of enabling technologies

Lifecycle Value Stream: Taking predictive maintenance as the example for edge analytics implementation, the lifecycle value stream of predictive maintenance has both managerial and technical implications. As far as the managerial perspective is concerned, the type includes the idea as well as the development and validation of a predictive maintenance strategy. After successful validation, the new consulting service is released. Each instantiation of the predictive maintenance strategy to a specific production process or industry represents an instance of that type. As far as the technical perspective is concerned, the type includes the idea as well as the development and testing of a unified information system for predictive maintenance which sets the basis for serial production. Each instantiation of the predictive maintenance information system to a specific piece of equipment, production process or industry, and to a specific legacy data system or installed sensor represents an instance of that type.

Core Design Principles

The following core design features & principles must be considered while designing any production or application or solution as given below:

- Flexible – it should be flexible enough to accommodate more product types

- Integrability – Open connectivity (such as OPC UA) with third-party systems through standard hardware and software as per corporate IT environment and safety.
- Compliant – System should be compliant to various standards (such as ISA-95) and regulations that are existing and evolving. There are prevailing local, regional, or international laws and regulations by various agencies to be compliant with in order to successfully deploy your application.
- Sustainable & Resilient Model – Investment in design and technology must ensure business sustainability model to ensure that investments are outcome-oriented, justify their business goals and are futuristic. Plus, these applications should be made resilient to any sudden change in business or market condition or calamity.
- Reliable and secure – The more the system is open to use, the more the risk of data loss, theft and cyber-attack. Reliability of system or data, security, availability of system and infrastructure are key to successful implementation. Probably, "vulnerabilities & changing security necessities" is another use case that prompts enterprises to embark on digital transformation journey. Safe and secure access to data is the benchmark of any service or technology provider. The role of CIOs, CTOs, system administrators, solution architects, and data base administrators has become more pivotal than before. Reliable use of cloud-based (private and public clouds), on-premises or hybrid model of infrastructure needed to deploy applications and process big data is a key consideration for a data driven business model.

- Responsive to Change – Changes in the Industry 4.0 landscape are imminent. However, responsiveness to changes is a must while designing the system. Changes could be in terms of communication protocols, operating systems, new databases, application upgrades, product types, user's access methods, mobility, user preferences, safety measures, third-party interfaces and so on.
- Scalable (ready for mass consumption) – Industry 4.0 typically relies on scalable Big Data or "web scale" data, which is based on high volumes and high velocity with millions of connected devices, scalable application deployment and concurrent users or mass consumption. Among the many technologies, DevOps is an evolving philosophy and framework that encourages faster, better scalable application development and faster release of new or revised software features or products to consumers.

  The practice of DevOps encourages smoother, continuous communication, collaboration, integration, visibility, and transparency between application development teams (Dev) and their IT operations team (Ops) counterparts. This closer relationship between "Dev" and "Ops" permeates every phase of the DevOps lifecycle: from initial software planning to code, build, test, and release phases and on to deployment, operations, and ongoing monitoring. This relationship propels a continuous customer feedback loop of further improvement, development, testing, and deployment. One result of these efforts can be the more rapid, continual release of necessary feature changes or additions.

- Modular – Applications must have modular design with global templates for local customisation for ease of diagnostics, debugging and deployment.
- Configurable – Many users find it tedious to deal with a new system developed by hi-tech consultants and engineers. Here comes configurability to rescue your design that allows users to amend and edit the existing codes, objects, and dashboards, among other things, any technologies that make it more user-friendly. Thus, actual users need not worry about complex design and components that work in the background to make it efficient and relevant at any stage, as they can pick or filter the data contents for their use.

Beside the above design considerations and architecture model for Manufacturing Industry, a design process must be followed that will guide you with 5-stages of Design Thinking – a well-known methodology described in a later section. Further, many applications within the manufacturing industry follow Cloud native Architecture (new term). Cloud native architecture and technologies are an approach to designing, constructing, and operating workloads that are built in the cloud and take full advantage of the cloud computing model as explained below.

Cloud-Native Architecture

Cloud native is about speed and agility. Business systems are evolving from enabling business capabilities to weapons of strategic transformation that accelerate business velocity and growth. It's imperative to get new ideas to market immediately. Typically, it's used by manufacturers using public or private or hybrid cloud platforms for many purposes such as cloud-based digital manufacturing, warranty support, or eCommerce applications.

The Cloud Native Computing Foundation provides the official definition.[28]

> *Cloud-native technologies empower organizations to build and run scalable applications in modern, dynamic environments such as public, private, and hybrid clouds. Containers, service meshes, microservices, immutable infrastructure, and declarative APIs exemplify this approach.*
>
> *These techniques enable loosely coupled systems that are resilient, manageable, and observable. Combined with robust automation, they allow engineers to make high-impact changes frequently and predictably with minimal toil.*

All the application functionalities, architecture diagrams and system design must be well documented keeping in mind the roles of key users, customers, maintenance, administration, and RUN-support activities involved later in the course of the Application Lifecycle.

For example, Volkswagen Group, Europe's largest car-maker, is to transform its automotive manufacturing and logistics processes. The Group built the Volkswagen Industrial Cloud on Amazon Web Services (AWS), which uses AWS Internet of Things (IoT) services to connect data from all machines, plants, and systems across more than 120 factory sites. The Volkswagen Industrial Cloud aims to yield a 30 percent increase in productivity, 30 percent decrease in factory costs, and €1 billion in supply chain savings. On AWS, the Group is expanding beyond manufacturing into ridesharing services, connected vehicles, and immersive, virtual car-shopping experiences to shape the future of mobility.

As per Dr Martin Hofmann – Group CIO – Volkswagen AG – 2019, they are working with AWS to move its 124 factory sites to a single architecture: the Volkswagen Industrial Cloud. As a next step, the Group will integrate Volkswagen's global network of more than 1,500 suppliers into the Industrial Cloud. The company uses AWS machine learning services—running algorithms using information gathered from sensors on the shop floor—and AWS Outposts.[29]

As per Amazon Press release dated March 27th, 2019; AWS will power the Volkswagen Industrial Cloud, automating all automobile manufacturing and logistics processes.[30]

The Five Stages of Product Design Thinking

The Hasso Plattner Institute of Design at Stanford (aka the d.school) describes design thinking as a five-stage process which is usually applicable to product design.

"Design thinking is a non-linear, iterative process that teams use to understand users, challenge assumptions, redefine problems and create innovative solutions to prototype and test. Involving five phases—Empathize, Define, Ideate, Prototype and Test—it is most useful to tackle problems that are ill-defined or unknown."[31] – Five Stages of Design Thinking (Source interaction-design.org, sighted on 31/08/2022)

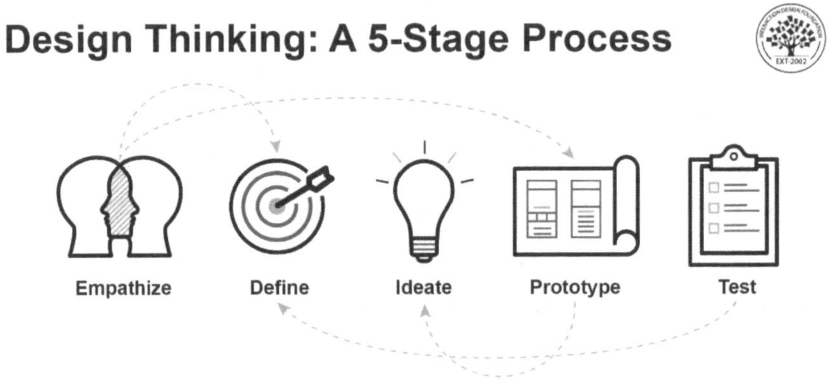

Figure 25: Design Thinking is a 5-Stage Process (Source: interaction-design.org)

Overall, you should understand that these stages are different modes which contribute to the entire design project, rather than sequential steps. Your goal throughout is to gain the deepest understanding of the users and what their ideal solution/product would be.

Communication Protocols

Big Data from Connected boardrooms, Connected Supplier Partners, Connected Designers, Connected Machines, Connected Products, and Connected Consumers is the new key to competitiveness in manufacturing.

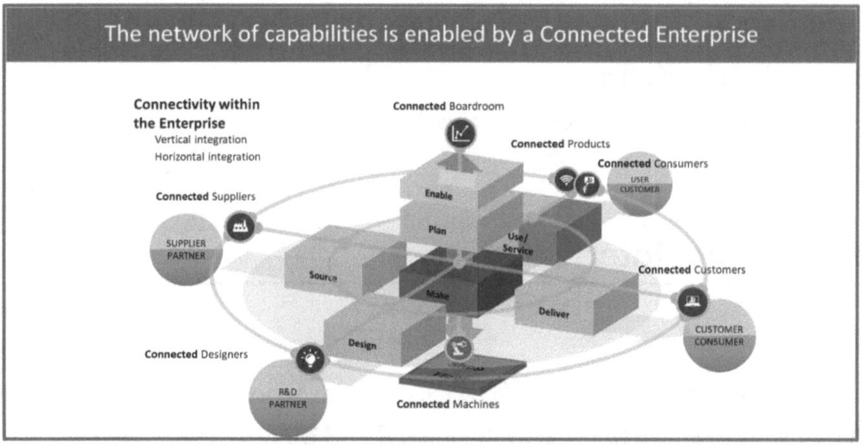

Figure 26: Connectivity Solution Diagram with an Enterprise (Source; Atos SE, 2021)

In Big Data, an enormous volume of data is used and this creates two main challenges. The first challenge is how to collect large volumes of data and the second challenge analysing the collected data. To overcome those challenges, you need a messaging system. Before we dive deep into Cyber-Physical System Use Cases and their benefits, it would be useful to touch upon some of the game-changing communication protocols and messaging systems that have enabled IT (Information Technology) & OT (Operational Technology) Convergence to expand so rapidly, including cloud-based IIoT and IoT platforms, such as:

- Proprietary interfaces
- OPC
- OPC UA
- Message Queuing Telemetry Transport (MQTT) and
- Advanced Message Queuing Protocol (AMQP)
- Apache Kafka
- Apache NiFi – S2S Protocol

Proprietary Interfaces

Especially in Industrial IoT (IIoT), this is the most common scenario. Machines provide many usually closed and incompatible protocols in a proprietary format.

Examples are Siemens S7, PROFINET, Modbus, Rockwell FactoryTalk Linx, or MelsecNet.

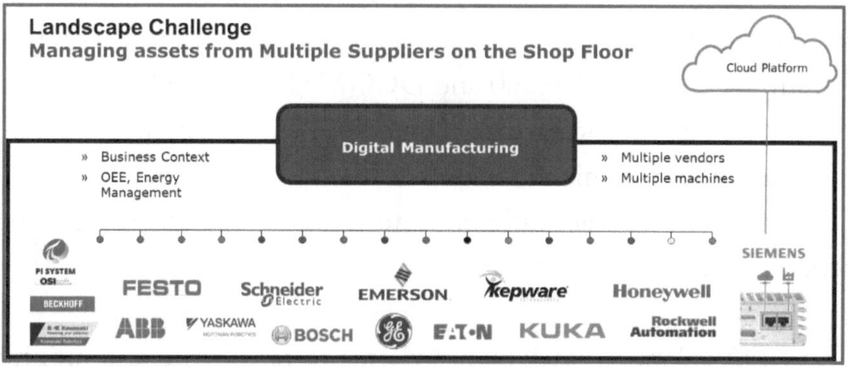

Figure 27: Digital manufacturing proprietary network landscape

Supervisory control and data acquisition (SCADA) and upper layer systems are often used to control and monitor multiple machines with disparate communication protocols.

OPC Protocol

Up until the 1990s, when you needed to integrate an industrial device into a Graphical User Interface (GUI), you first had to figure out which protocol to use. Although widely used protocols already existed, the devices to be integrated often used different ones, making the developer's job very expensive.

Since the graphical interfaces were, in all cases, developed in the Windows environment, it was decided to simplify life by introducing a new communication platform, used today, called OPC (Open Platform Communication). It is based on Microsoft's proprietary COM (Component Object Model) and DCOM (Distributed Component Object Model) interfaces which allow communication between running processes. The transmission of data between client and server in OPC Classic is done through different interfaces defined according to the purpose, and the main ones are:

- DA (Data Access): allows clients to read, write and monitor variables in real time
- HA (Historical Access): used to access already-stored data
- A&E (Alarms and Events): manages alarms and event notifications.

OPC UA Protocol

Over the years, the use of Unix devices in the industry quickly led to the need to extend the functionality of this platform that was otherwise only applicable to Windows systems.

In addition to this, OPC Classic was a poor tool for data exchange, allowing only primitive data types to be transferred. Around 2008, the second version called OPC Unified Architecture (OPC UA) was developed which, among many other things, offers:

- Full compatibility with the most popular operating systems (Windows, Linux and Mac)
- Support for Internet protocols for distributed communication
- Description of complex data models with an approach very similar to the OOP paradigm
- Possibility to expose remote methods
- Scalability

We can say that OPC UA is based on 3 fundamental pillars: the transport mechanism, the management of calls to remote methods and the data description model. Regarding the first point, OPC UA supports both a TCP-based binary protocol that guarantees high performance in transmission, and two widely used web communication protocols such as HTTPS and SOAP.

This extended the previous version, adding the possibility to also communicate via Internet between different devices. The possibility, then, to call remote methods, makes the protocol an incredible tool to expand the potential of devices such as Programmable Logic Controllers (PLCs). For those who don't know, the PLC is a programmable controller widely used for real-time control of machines in industrial automation due to its reliability and stability. On the flip side, PLC systems are very crude and don't offer the same possibilities and connectivity that a PC can provide.

To give an example, if we wanted a PLC to write to a database, we would have to use a PC as a mediator and then implement complex (and difficult to manage) methods to allow communication between the PLC and the PC. But, as we speak, this concept is changing for many PLC manufacturers who are incorporating new features. OPC UA, in this sense, extends the features of PLCs, making it simple and immediate to call remote methods.

Thus, a PC or PLC with an OPC Server can expose more or less sophisticated methods, such as database read/write functions, that can be evoked by the machine controllers when needed.

Talking finally about Data-Modelling, OPC UA offers a complete modelling system that allows one to describe data of variable complexity. This is undoubtedly the most important strength of the whole platform as it is possible to create systems in which data exchange takes place with scalability and a completely decentralized infrastructure.

In this way, for example, instead of leaving the PLC the burden of collecting data from dummy sensors, and then organizing the information and presenting it to the upper layers, you can let the sensors themselves send complex data directly, leaving the PLC with the sole task of controlling the progress of production in real time. In order to do this, it is essential to have a communication protocol that can handle complex data. This is where OPC UA comes in and changes the game, allowing information to be represented in units called nodes. Like an OOP class, each node, once described, can be instantiated into different objects with attributes and methods giving rise to models of increasing complexity. A single OPC UA router product can act both as Subscriber and Publisher at same time.

inray Industriesoftware GmbH explains, in their article *The OPC UA Pub/Sub Plug-in*,

> "*the true benefits of OPC UA Pub/Sub Plug-in feature: "With the OPC UA Pub/Sub Plug-in, the OPC Router offers a new type of communication for OPC UA data. The Pub/Sub communication works based on the Publish-and-Subscribe principle. The OPC Router is integrated into this communication. It is able to act as a Publisher (sender) and as a Subscriber (receiver) and can therefore exchange and receive information. The decoupling of sender and receiver leads to much simpler communication, as n-to-m connections are possible.*"[32]

This type of innovation leads to endless possibilities of communication among devices and cloud as illustrated below.[33]

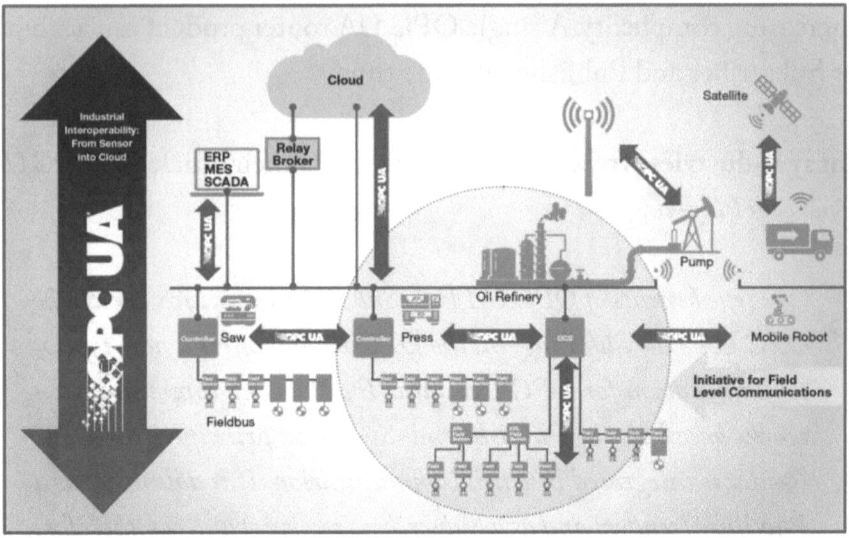

Figure 28: OPC UA for Field Level Communication (Source: https://opcconnect.opcfoundation.org)[34]

PLC4X

As an Apache framework, it provides a unified API by implementing drivers (similar to JDBC for relational databases) for communicating with most industrial controllers in the protocols they already understand. No license costs or hardware modifications are required.

MQTT & AMQP Protocols

In early 2018, Part 14 of the OPC UA Standard, aka "Publish-Subscribe" or "PubSub" was released, helping to encourage further interaction between OPC UA and communications protocols such as MQTT. According to the OPC Foundation,

> *"PubSub enables the use of OPC UA directly over the Internet by utilizing popular data transports like MQTT (Message Queuing Telemetry Transport) and AMQP (Advanced Message Queuing Protocol) while retaining its key OPC UA end-to-end security and standardized data modelling advantages".*[35]

The standard has also helped further communication abilities within IoT devices such as gateways. IoT gateways provide a bridge between an on-premises communications network and a cloud-based communications network. Sometimes referred to as "Edge devices", IoT gateways provide data connectivity to sensors and end-devices, completely on-premises. They also include a communication path that provides data connectivity between itself and the cloud. Today's intelligent, programmable devices perform in-depth and complex processing on IoT data, acting as full-fledged computing platforms running modern operating systems, for example, Linux or Windows.

Such systems are sometimes also called "intelligent gateways," due to the increase in processing power, memory and storage, as well as intelligent "middleware" encompassing key features such as protocol translators, data buffering, edge analytics, on-premises visualization and cloud enablement, which support emerging communication protocols including MQTT and AMQP for publishing data to the cloud.

In the case of the process, factory and building automation industries, the Southbound communications protocol is typically an industry standard protocol, such as OPC UA. It could also involve BACnet, Modbus, SNMP or Web services, or could also be a proprietary communications protocol. The Northbound communication path needs to be highly secure and is often based on a publish/subscribe ("pub/sub") mechanism. Regarding the trade-offs of MQTT, consider the pros and cons:

Pros

- Widely adopted
- Lightweight
- Has a simple API
- Built for poor connectivity and high latency scenarios
- Supports many client connections (tens of thousands per MQTT server)

Cons

- Just queuing, not stream processing
- Inability to handle usage surges (no buffering)

- Most MQTT brokers don't support high scalability
- Asynchronous processing (often offline for long time)
- Lacking a good integration with the rest of the enterprise
- Single infrastructure (typically somewhere at the edge)
- Inability to reprocess events

These trade-offs show that MQTT is built for IoT scenarios but requires help when it comes to integrating into the enterprise architecture of a company. This is where the event streaming platform Apache Kafka and its ecosystem come into play.

Apache Kafka

Kafka is an out-of-the-box Connect interface and integrates with hundreds of event sources and event sinks including Postgres, JMS, Elasticsearch, AWS S3, and more. It has in-built stream processing of events with joins, aggregations, filters, transformations, and more, using event-time and exactly-once processing features. It allows client libraries to read, write, and process streams of events in a vast array of programming languages. With a large ecosystem of open-source tools to leverage a vast array of community-driven tooling, it's a widely accepted interface for big data collection such as alarms, industrial data, events and more. The following diagram illustrates the end-to-end Kafka Data Pipeline.

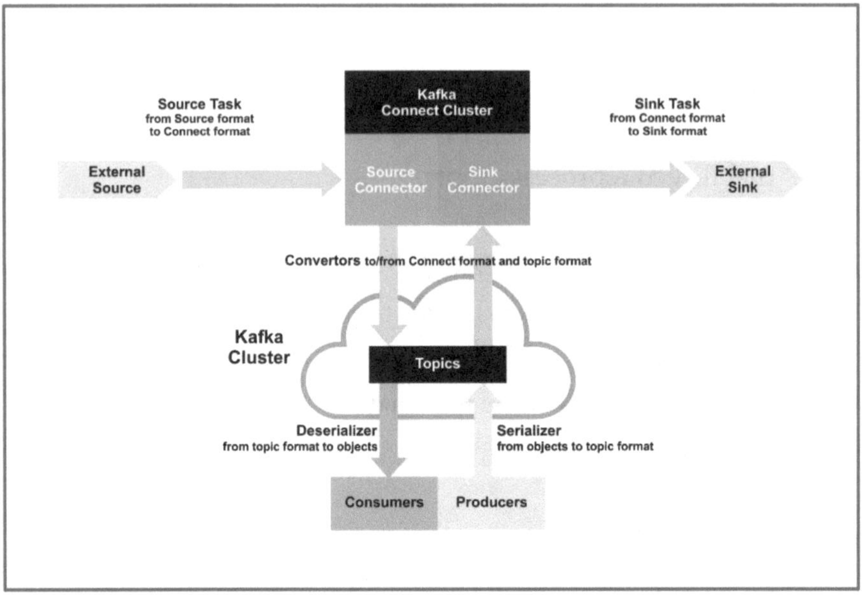

Figure 29: End-to-end Kafka Data Pipeline (Source: www.instaclustr.com)[36]

Kafka is considered as a perfect fit for event streaming. Event streaming is the digital equivalent of the human body's central nervous system. It is the technological foundation for the 'always-on' world where businesses are increasingly software-defined and automated, and where the user of software is more software. For example, Kafka is a potential messaging and integration platform for Spark Streaming. Kafka act as the central hub for real-time streams of data and are processed using complex algorithms. Once the data is processed, Spark Streaming could be publishing results into yet another Kafka topic or store in HDFS, databases or dashboards.

Technically speaking, event streaming is the practice of capturing data in real time from event sources like databases, sensors, mobile devices, cloud services, and software applications in the form of streams of events; storing these event streams durably for later retrieval; manipulating, processing, and reacting to the event streams in real time as well as retrospectively and routing the event streams to different destination technologies as needed. Event streaming thus ensures a continuous flow and interpretation of data so that the right information is at the right place at the right time.

Apache Kafka is a message broker with the focus on storing massive amounts of data, and allowing consumption in real time or later utilization. Kafka uses its own network protocol. We need to store massive amounts of messages, ensure batch processing, so we need Kafka. Kafka itself has no built-in message priority, security is poor and the heavy protocol makes it difficult to use.

MQTT does not seem to have the ability to handle very high throughput from sensors so again we have to turn to Kafka which can handle high throughput. If we use things like Arduino Yun, as it's one of processors that runs OpenWRT Linux distribution, we can simply port Kafka to this OS and run the Kafka client to push data to the Apache Kafka server.

Some of the Kafka-native options for MQTT integration beyond Kafka client APIs like Java, Python, .NET, and C/C++ are:

- Kafka Connect source and sink connectors, which integrate with MQTT brokers in both directions
- Confluent MQTT Proxy, which ingests data from IoT devices without needing a MQTT broker
- Confluent REST Proxy for a simple but powerful HTTP-based integration

Most scenarios require a reliable, scalable, and secure end-to-end integration that enables bidirectional communication and data processing in real time.

Apache NiFi – S2S Protocol

NiFi was built to automate the flow of data between systems. While the term 'dataflow' is used in a variety of contexts, we use it here to mean the automated and managed flow of information between systems. The preferred communication protocol between NiFi instances is the NiFi Site-to-Site (S2S) Protocol. S2S makes it easy to transfer data from one NiFi instance to another easily, efficiently, and securely. NiFi client libraries can be easily built and bundled into other applications or devices to communicate back to NiFi via S2S. Both the socket-based protocol and HTTP(S) protocol are supported in S2S as the underlying transport protocol, making it possible to embed a proxy server into the S2S communication.

NiFi is designed to scale out clustering many nodes together as described above. If a single node is provisioned and configured to handle hundreds of MB per second, then a modest cluster could be configured to handle GB per second. This then brings about interesting challenges of load-balancing and fail-over between NiFi and the systems from which it gets data. Use of asynchronous queuing-based protocols like messaging services, Kafka, etc., can help. Use of NiFi's 'site-to-site' feature is also very effective as it is a protocol that allows NiFi and a client (including another NiFi cluster) to talk to each other, share information about loading, and exchange data on specific authorized ports.

Manufacturing Execution System (MES/ MOM) (Level -3)

An Optimized Manufacturing Operation gives intelligent visibility and control in real time rather than hours, days or months. Decision-making power has never been so quick. The gaps between supply-chain, logistics, manufacturing capability, and market demand can be minimized in a dramatic manner. Mass-customization to meet individual customer demands can be met easily and that leads the eCommerce engine to work efficiently.

MES is a specialist class of production-oriented software that manages, monitors, and synchronizes the execution of real-time, physical processes involved in transforming raw materials into intermediate and/or finished goods. These systems coordinate this execution of work orders with production scheduling and enterprise-level systems like ERP and product life cycle management. MES applications also provide feedback on process performance, and support component and material-level traceability, genealogy, and integration with process history, where required.

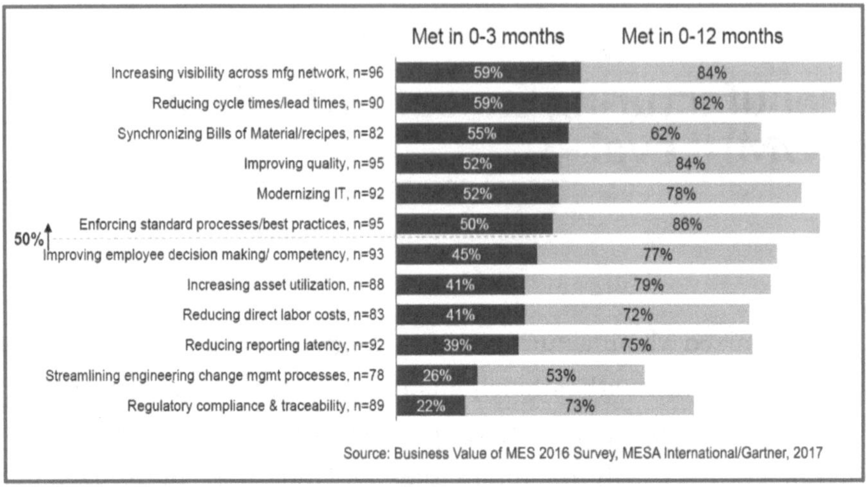

Figure 30: Business Value of MES Implementation Survey (Source: MESA International/Gartner, 2017)

To advise on the execution of MES systems and address their growing complexity, MESA, the Manufacturing Enterprise Solution Association, was created in the 1990s. Their model is perhaps the most widely used in the industry as it dates back to 1997, when MESA formally defined the scope of MES through 11 core functions called the MESA-11 model.[37]

Manufacturing execution functionality is composed of the 11 critical capability areas outlined below. Even though the MES System was not new when the Industry 4.0 arrived, the use cases of MES have grown exponentially. Creativity plays an important role in adopting the Industry 4.0 platform and that takes it to the whole new level of realization of benefits. For example, industry has moved from an on-premises MES system to a hybrid or cloud-based MES System to achieve greater heights from the same concept. How these reflect the trends in this market and area of technology is explained further below.

1. Production Management, Execution/Workflow

This means coordinating the execution of material movement and task/recipe execution with prompted operator interaction. Also, the auto-selection and download of appropriate control recipes or discrete tasks based on unit operations and/or recipe steps, and configurable workflow tools to enable these tasks.

Production execution is one of three key components of MES, a capability that separates MES from systems that only monitor production or only collect production data for analysis. The importance of production execution varies from use case to use case. For example, in some continuous process manufacturing scenarios, production execution resides in the process control system, so MES-related production execution is not as important.

Manufacturing continues to move toward single-piece flow, personalized products and small batch production, so rapid configuration changes are the rule, not the exception. Some MES products also use these generic workflow tools to create workflow models outside of the production environment for tasks such as maintenance management, field service or nonconformance/corrective action management.

2. In-Process Quality Management

This provides the ability to import critical quality requirements from external systems and ensure that those quality requirements are met, and also the ability to define and/or manage the corrective action process when there are deviations.

Quality management at the MES level is not a specific requirement of an MES system. In some cases, an end-to-end quality management system would take the place of MES in-process quality management. In most cases, there will be some sharing of functionality between MES and the quality management system (QMS), and this will present additional integration challenges.

3. Data Management/Data Collection

This provides the ability to access and manage data collected via end-user interaction and/or process historians for time series and transactional operational data.

Regardless of use case, midsize data management (MDM) can and will extend to production data. MES systems must have the capability of accessing, managing and sometimes contextualizing transactional and time-series data in an integrated environment. Planning for these data management tasks via creation of global implementation templates and metrics definitions is key to enterprise MES implementation.

4. Regulatory Compliance/Track-Trace/Automated Routing

Compliance with industry-specific regulations requires the ability to create, manage and control transactional records for parts, components, ingredients, and processes, and track them at the batch, lot and/or serial number level.

Life sciences, food/beverages, and aerospace and defence industries have stringent regulatory compliance requirements, but the need for compliance/track and trace is not unique to those industries. Relationships between raw materials/ingredients, component parts, assemblies, co-products, by-products across the supply network and life cycle need to be maintained, and much of the complexity is created, and must be managed, in manufacturing.

Automated routing comes with in-built decision-making algorithms with the help of pre-defined business rules for routing the products on the shop floor, in delivery or during shipment and has become a game changer in the era of Industry 4.0. Not only manufacturers but large online shipping and delivery companies have made it their competitive advantage and a barrier for others to foray into the high-volume shipping market.

For example, we have the hive-grid machine used in a British online-only supermarket that's made a name for itself in recent years designing highly automated warehouses and selling the tech to other grocery chains. When fully up and running, it will be its most advanced yet, processing 3.5 million items or around 65,000 orders every week. It's also a perfect example of the wave of automation slowly hitting countries around the world. The automated tasks being undertaken by bots are so basic that they're best described by simple verbs — "lifting," "moving," "sorting" — and that means they exist in various forms in a range of industries. As explained by Ocado's chief technology officer, Paul Clarke – the goal is to "disrupt itself" ; to continually upgrade its technology so it can't be overtaken by competitors. In the company's older, now-outdated warehouses (which Clarke claims are still some of the most advanced in the world), groceries are processed in a roughly linear fashion. Deliveries are unpacked into crates, crates are placed onto conveyors and conveyors carry the crates to shelves, where human "pickers" take what they need to fill customers' orders. The new paradigm, though, is all about using space as efficiently as possible. Items are still placed in crates, but those crates are now stored in huge stacks, up to 17 boxes high. Their position in this stack seems to be at random — a box of razors next to cod fillets, for example — but it's algorithmically decided with frequently accessed items placed on the top and rarer purchases near the bottom. On top of this hoard, the robots do their work.[38]

5. Analytics/Reporting/Performance

Analytics, reporting and performance make production data visible to business leadership across the manufacturing network, from plant to plant and across the shop floor.

There is built-in reporting of standard key performance indicators (KPIs). Clean, easy-to-understand and device-independent dashboards are available for production monitoring and processing KPIs. Dashboard configuration and ad hoc report generation tools are readily available and simple to use.

6. Production Equipment Integration

This demonstrates the ability to integrate with production equipment (e.g., process historians, robots, PLCs, SCADA systems, edge devices and data collection systems).

Many COTS software providers have libraries for various industries, processes, domains and equipment monitoring available at no cost along with trial licenses. Also, they have sample files available for Cloud Connect, Web Interface, WebUX and WebNavigation. For example, Siemens provides many SIMATIC WinCC V7.5 SP2 Demo Projects that are ready to configure and use by System Integrators.[39]

This may include specific equipment connection libraries and/or toolkits for the development of integration with historians, SCADA systems and/or equipment. Scores also reflect the openness of these integrations. For example, does the vendor integrate with multivendor production environments equally, or are there limitations for some vendor production equipment integrations?

7. Enterprise Integration Architecture

This demonstrates the ability to integrate with other enterprise systems for bill of materials (BOM)/recipe management, materials management, inventory, order status, and completions for systems such as ERP, PLM and supply chain management. We are looking for sophisticated process and data integration such as Master Data Integration:

- Order Download
- Batch creation with Related Characteristics
- Production Confirmation
- Scrap/Waste Declaration

This includes the availability of vendor-certified integrations, APIs, software development kits (SDKs), microservices or other technologies for deployment of these integrations. Scores also reflect whether the MES vendor provides a full integration solution or a partial solution (its "half of the bridge"). In the latter, the end-user customer or another vendor may be required to do an equal or greater amount of work to make the integration operable. This is a key requirement, and a weakness, in some MES solutions.

8. Usability/User Experience

User experience (UX)/user interface includes responsive UI design, workflow management tools, visualization capabilities, and UX differentiated between operators, supervisors, back-office and managerial roles, as well as user-specific profiles (language, screen layout, etc.).

Usability also includes support for multiple modes of interaction, such as mobile devices, voice and AR.

Gartner research finds that users are now placing greater emphasis on the usability of MES across different classes of employees. Issues like personalization, user-level localization, low-code/no-code app development, user onboarding, and capabilities to make the system appeal better to new and different types of users are important. Look and feel, navigation and process streamlining are also considered.

9. Deployment Experience and Options

This capability includes both the availability and ease of use of standard tools to aid implementation and the availability of industry-specific global implementation templates and best practices, supported by a variety of robust, production-ready options for hosting applications and support. As stated by Sutter, Burr (2017) in his eBook: Teaching an elephant to dance.[40]

> *"Applications have moved outside the IT department. There's a truism that all companies are now software companies, and the ability to rapidly provide new services and new functionality to customers is one of the key competitive differentiators a company can offer. IT agility is a stone that startup Davids can use to unseat massive Goliaths."*

Templates and capabilities are available in localization, hosting and licensing options. The size and quality of the implementation partner network are also part of the focus. Additionally, licensing, hosting and support options are plentiful and robust, including on-premises, private cloud, public cloud and hybrid approaches (including edge devices).

10. Ease of System Upgrade

This critical capability measures the quality of system upgrade tools, speed of upgrade and ability to migrate local customizations and configurations.

By definition, MES systems are expected to have extremely high levels of availability. This critical capability measures the migration tools and processes, how well upgrades are performed and how disruption to manufacturing production is minimized. It is not uncommon to find production MES systems where upgrades have never been done for fear of extended production downtime or instability.

11. MES/MOM Architectural Innovation

MES architectural innovation means the adoption of innovative core technologies, tools and processes that improve the state of the art for this market.

Architectural innovation for MES / MOM assumes full service-oriented architecture (SOA) with broad integration scenarios, the creation of composite applications, and support of enterprise process templates and orchestrated workflows as a starting point. Cloud connectivity is a modern solution being adopted to maintain a tight connection between MES, ERP, and the supply chain.

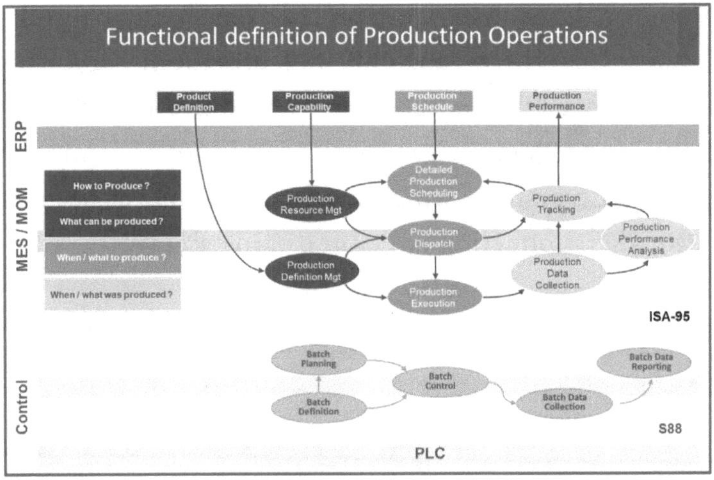

Figure 31: Functional Definition of Production System and Data Flow (Source: www.isa.org)[41]

In modern production environments, more equipment and machines are becoming connected—and need to be connected to an MES to provide real-time actionable data. This capability is rated based not only on the technology supported but the length of time it has been in production use at manufacturing sites. Solutions are also assessed by the level of cloud penetration, maturity and type of cloud enablement (virtualization versus native cloud), use of microservices-based architectures, and low code/no-code app development. The assessment also includes support for wearables, augmented reality (AR)/virtual reality (VR) and new approaches to voice. Solutions are also assessed for new approaches to how existing technologies like mobile devices may be handled.

For instance, in a typical automotive manufacturing plant, there may be thousands of OT devices—anything from cameras and PCs to robots and various controllers. That plant can also have more than 50 manufacturing execution systems (MES) applications and because the ecosystem is connected, that means more than 100,000 connections between the OT and IT layers of the software stack.

Various Smart Factory solutions built around Industry 4.0 have become core to business advantage for manufacturers to improve several performance factors (KPIs) as highlighted in the following sections.

Business Use Cases

Every business case viability is based on the justification for the investment for a value in return. Making a successful business case for your new project is the winning way to ensure a good beginning for your business. How often have you been asked to "work the numbers" and provide a basis for a compelling project? Projects with a solid business case return value to the business, to their sponsors, and to the stakeholders and customers. Meeting scope, staying within budget, and getting things done on time are the tactical elements that deliver the value.

Intelligent Manufacturing projects return certain value in terms of implementation of KPIs monitored to improve productivity. Every KPI needs a clearly defined goal. What are you trying to achieve? Is it even something that can be achieved? What is the time period within which you wish to achieve this goal? Be specific and set up parameters that clearly define your goal and relate to respective hierarchical nodes. This goal needs to be something that can be numerically defined (quantitative not qualitative) for a plant, a team, a process, or a logical entity.

It is very important that you can objectively measure your manufacturing operations toward the goal. This means collecting and interpreting data; which bring us to the next criteria.

Data, data, data. There must be a clearly defined data source with a strict procedure for how the data are measured or collected. An automated data collection system increases the accuracy and reliability of information. There should be nothing left to manual interpretation or intervention here, giving a single truth across the organisation.

Reporting your data is just as important as collecting it. Different manufacturing KPIs will have different reporting frequencies. Typically, reporting should happen on a weekly or monthly basis, or on event/alert basis. In the following sections, we will discuss some of the critical KPIs or functionalities, which provide a competitive edge to the manufacturing industry, that can be achieved using an Intelligent Manufacturing system.

Productivity Suite

Productivity Suite consists of several KPIs meant to improve productivity, reduce losses, and provide end-to-end insight into Manufacturing Operations. Close to the heart of key stakeholders, it's the core technology investment to improve efficiency and reduce bottlenecks, and other parameters online in real time can give you multi-fold returns. Some of the KPIs in Productivity Suites are listed below.

Agile Production Planning and Scheduling

An Agile Production Planning and Scheduling system allows agile rescheduling upon the occurrence of unexpected events. It executes schedule-based planning in the sense that scheduling is tightly connected with net-requirements planning and capacity-requirements planning. In such an agile system, provision is made to allow changing various parameters dynamically as highlighted below.

- Auto Scheduling for Subassembly /Child component based on final assembly and the cycle time
- Agile Sequencing of the shop orders based on the man-machine availability for a day/shift
- Re-adjusting of Production Model Types and volumes
- Agile re-sizing of Production batches based on running constraints or priority
- Re-adjusting of shift-timings and durations
- Re-adjusting of break-timings and durations

Advanced Manufacturing Execution systems (MES) are well connected in real time with third-party interfaces, configurable and flexible to facilitate agile production planning and scheduling.

Thus, it allows greater flexibility of operations, cost saving, reduces idle time, reduces inventory, and allows for optimum utilisation of resources and an increase in productivity or revenue.

Overall Equipment Effectiveness (OEE)

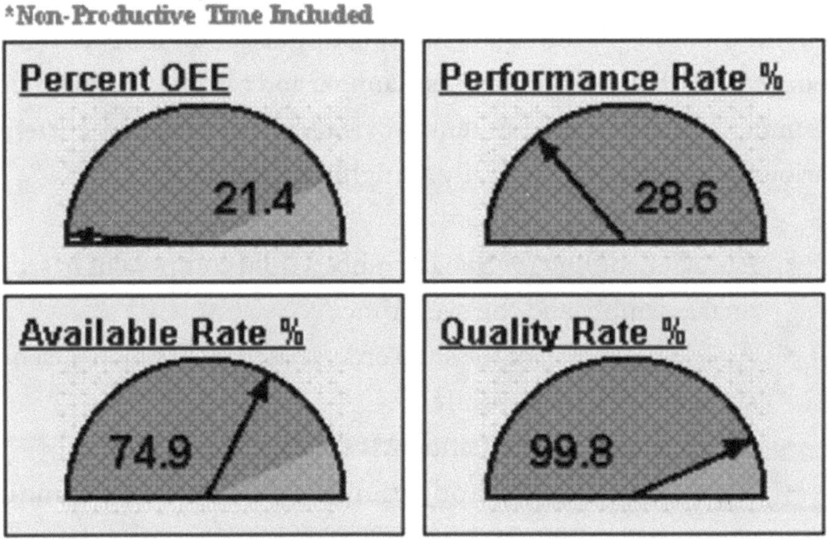

Often, productivity and efficiency of equipment and plant are the key parameters to watch at all the levels – be it at enterprise level or by plant heads or operational heads. Not only does it measure the overall plant performance, but many times, it's also attached to the team's performance and incentives. It's equally important that a single truth prevails at all the levels. So, the whole team and management can easily understand the methodology to measure it and monitor the improvement over a period of time.

Many terminologies are prevalent in industry to monitor the performance KPIs of equipment or plants – such as Overall Equipment Effectiveness (OEE) or Operational Performance Rate (OPR), etc. As explained below, the OEE contains several KPIs together to monitor Expected vs actual Utilization of equipment.

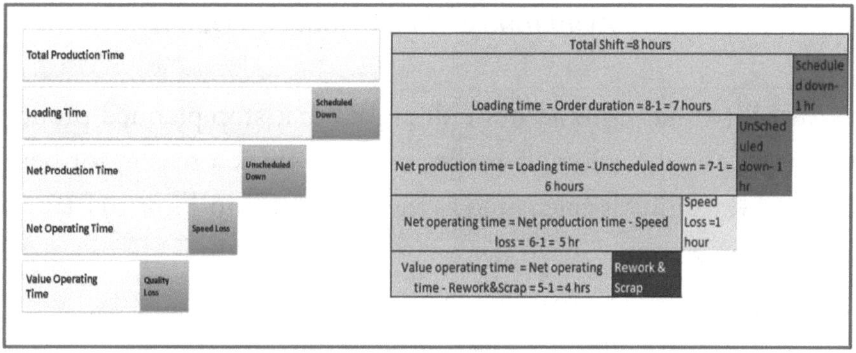

Figure 33: OEE Calculations demonstration

This key performance indicator is considered the gold standard for measuring manufacturing productivity. The simplest way to calculate OEE is as the ratio of Fully Productive Time to Planned Productive Time. Fully Productive Time is just another way of saying manufacturing only good parts as fast as possible (Ideal Cycle Time) with no Stop Time. Hence the calculation is:

$$\text{OEE} = \left(\frac{Good\ Count * \ Ideal\ Cycle\ Time}{Planned\ Production\ Time}\right)$$

The higher your OEE, the more effective your equipment is. A score of 100 percent means that you are manufacturing 100 percent of the time, at 100 percent capacity, at a 100 percent yield (no defective parts) and can be calculated as below.

$$\text{OEE} = Availability * Performance * Quality$$

Availability takes into account all events that stop planned production long enough where it makes sense to track a reason for being down. Availability is calculated as the ration of Run Time to Planned Production Time:

$$\text{Availability} = \left(\frac{Run\ Time}{Planned\ Production\ Time}\right)$$

Whereas, Run Time = Planned Production Time – Stop Time

Performance, also referred to as Operational Performance Ratio (OPR), considers anything that causes the manufacturing process to run at less than the maximum possible speed when it is running (including both Slow Cycles and Small Stops). Performance is the ratio of Net Run Time to Run Time. It is calculated as:

$$\text{Performance} = \left(\frac{Ideal\ Cycle\ Time * Total\ Count}{Run\ Time}\right)$$

Ideal Cycle Time is the fastest time that your process can achieve in optimal circumstances. Therefore, when it is multiplied by Total Count, the result is Net Run Time (the fastest time to manufacture the parts:

Since the rate is the reciprocal of time, performance can also be calculated mathematically as:

$$\text{Performance} = \left(\frac{\frac{Total\ Count}{Run\ Rate}}{Ideal\ Run\ Rate}\right)$$

Performance should never be greater than 100%. If it is, that usually indicates that the Ideal Cycle Time is set incorrectly.

Quality losses often get missed out on in-line production systems for dynamic calculations of OEE. Thus, OEE sometimes is calculated without in-line quality results. In many cases, quality losses are measured on a sample basis on the finished goods, whereas some production lines allow in-line quality testing and rejection of parts/batch based on automated/ manual testing.

Although it is never easy to segregate the reasons for the losses and inefficiencies caused by several planned and un-planned downtimes, poor quality, starved stations, bottlenecks, or slowness of the operation, it still pays to monitor them live and then take corrective actions in a near-instance environment instead of analysing the reasons like a post-mortem.

With the real-time bi-directional connectivity between shop floor and board room across the enterprise and corporate level, one can take corrective action using vertical integration of systems, visualization and analytical tools available along with the help of tools available under Industry 4.0.

Shop-floor devices must be configured in such a way that they continuously broadcast data about the status of productivity. Some of the typical shop-floor events that count towards line stoppage or production loss are listed below:

- Operator Stops, or E-Stops
- Safety Gates or Light Curtain Faults
- No Power, Air or Energy
- Breakdowns or Miscellaneous Stops
- Empty Carrier/ No-part /Test Unit/No-Ops Unit
- Operating Loss
- Quality loss – Non Conformance (NC)
- Loss of Time Cycle
- Losses of Product/ process
- Managed Stops
- Blocked Downstream
- Starving for Upstream Products

- Waiting for Material or Sub-components
- Planned Downtime – Scheduled Maintenance or Break-times

Comparing productivity among various types of equipment, production lines, shops and sites gives a quick insight on which to make informed decisions at various levels to reschedule production or remove bottlenecks and to identify short-term and long-term fixes to increase the productivity levels.

Throughput

This is probably one of the most fundamental KPIs for the manufacturing industry while also, arguably, one of the most important. The Throughput KPI measures the production capabilities of a machine, operator workstation, line, sub-area or plant; also known as how much they can produce over a specified time.

$$\text{Throughput} \quad = \left(\frac{Number\ of\ Units\ Produced}{Time\ (hour\ or\ day)^{\#}} \right)$$

Time could be of two types, namely the net production time or the clock time.

Thus, two different throughput rates can be reported for the same amount of production and the same duration depending upon the purpose.

Cycle Time

The cycle time KPI is a very simple in nature, but that doesn't mean it can't be manipulated to be a very powerful tool. In the manufacturing industry, cycle time is the average amount of time it takes to produce a product or part. Simply stated, a cycle time

Cycle Time = *Process End Time – Process Start Time*

metric can be used to measure the time it takes to manufacture a complete product, an individual part of a product, or the time it takes to deliver a product to a customer.

This could be a cycle time of a manual, semi-auto or automatic operation, and the efficiency can be determined for the operator /team, the machine or the process. However, the operator/ team may not be a part of the machine but may keep on changing depending which shift is operating or working on a particular station or operations. In a typical advance & automated system, it should be possible to link the operator/ team with the workstation(s) during each shift to measure the efficiency per operator/team. An incentive can be triggered automatically from the system if a certain target of efficiency is achieved by an individual operator/ team for a given workstation/ operation.

In case of an inefficient operation, it is possible to find the root cause using advance analytical tools for a given period of operation to allow for further corrective actions.

Hence, cycle time can be used to monitor the overall efficiency of a manufacturing process on a macro scale, as well as to determine the inefficiencies on a micro scale.

Changeover Time

At the most basic level, changeover time represents the amount of time required to switch from one task to another. Typically, in manufacturing, it represents the amount of time lost from switching a production line from one product to another. However, it can also represent the amount of time lost during a shift change.

Changeover Time = *Net Available Time – Production Time*

Some manufacturers factor this time as net availability loss and not as performance loss, thus it is not considered as downtime of equipment.

Takt Time

This is a very useful manufacturing KPI when scheduling production orders or deciding whether to take an order from a client. Takt time is the maximum permissible amount of time that can be spent manufacturing a product while still meeting a client's deadline. For those who are curious, Takt stands for "taktzeit," a German word meaning “cycle time” . While very similar in nature, this is not to be confused with the cycle time KPI.

$$\text{Takt Time} \quad = \left(\frac{Net\ Available\ Time}{Customer's\ Daily\ Demand} \right)$$

Manufacturers outsource the part of production or look for alternate ways to meet the demand if the Takt time is not achievable within their facility. Sometimes, even if a maximum number of lines can meet the desired Takt time, certain equipment or lines may be the bottlenecks. As illustrated below, an advance MES system can monitor the bottleneck and Takt time of the production line in real time.

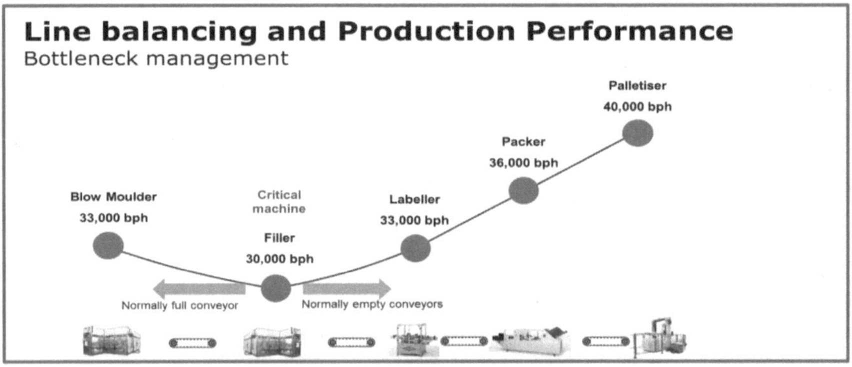

Figure 34: Bottleneck Management Solution

As a practical rule, it is always the slowest equipment or line that is the deciding factor to measure the overall achievable Takt time.

Track and Trace

It was never as easy to track and trace products or parts without the help of Industry 4.0 connectivity and tools while on the shop floor or on the road. It's easier to know the location, status (operations or quality NC, etc.), counts, speed, weight, SKU code/ID and other useful parameters of a part at any given time with the help sensors, Wi-Fi, Bluetooth, and 4G/5G connectivity.

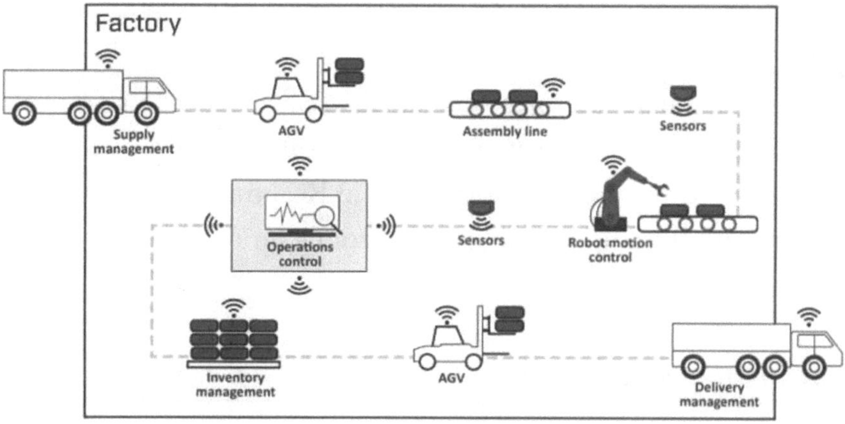

Figure 35: Advance Manufacturing value chain

As an extension of Digital platforms such as MES, SCM, or WMS, tracking and tracing supports Agile Production planning, operations, and distribution to gain on productivity metrices and customer satisfaction levels.

Inventory Turns

This is a measure of how many times inventory is sold over a specific time period and helps indicate resource effectiveness. Low ratio numbers indicate poor sales and excessive inventory, while high ratio numbers represent strong sales or insufficient inventory.

$$\text{Inventory Turns} \quad = \left(\frac{Cost\ of\ Goods\ Sold}{Average\ Inventory}\right)$$

A well-configured and integrated ERP, MES and WMS system can dynamically monitor the stock levels and sales accomplished. Thus, it's easier to know the Inventory Turns in Real time than in weeks or months with the help of Intelligent Manufacturing systems.

Return on Assets (ROA)

This being a financial metric, it plays as important a role as other manufacturing metrics. You can't have a business if you aren't making money. This metric evaluates how well your business is making use of its assets (money). It is the annual net income divided by total assets (fixed assets + working capital).

$$\text{Return on Assets (ROA)} \quad = \left(\frac{Net\ Income}{Average\ Total\ Cost}\right)$$

A company's working capital consists of its current assets (not including cash) minus its current liabilities (not including funded debts—such as mortgages and income taxes). Current liabilities consist of the orders pending to be fulfilled with payment received.

Management of working capital for manufacturing companies includes optimizing your account receivables and inventory processes. It ensures that a business has the resources and capital to grow a healthy operation and meet the current liabilities. As Stephan Bush states in an article in CHRON newsletter,

> "*Working capital is near the top of the list among business challenges currently faced by manufacturing companies.*"[42]

A well-designed intelligent manufacturing system may address this issue and ensures that inventory and orders pending are accounted for in the system at any given time. Below is the list of items typically exchanged between an ERP system and an MES system.

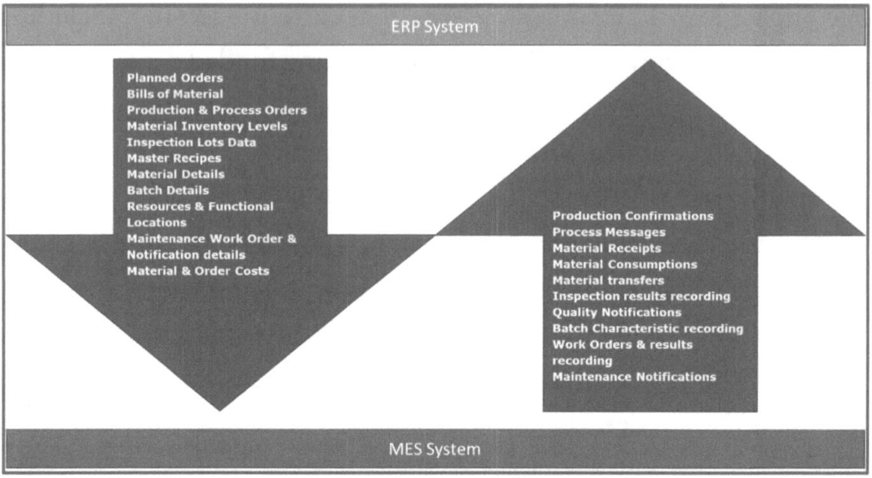

Figure 36: Typical ERP-MES Interface topics

A negative working capital prompts the stakeholder to take appropriate actions before they run into near- or long-term liquidity issues. An automated and integrated ERP- MES system can be configured to collect the real-time data for consumption of material per batch or final product or additional inventory (money) needed to complete the planned order.

Downtime to Operating Time

This manufacturing metric can be used to measure the effectiveness of machinery maintenance and the machine itself. With effective preventative maintenance, the amount of downtime can be reduced, creating a more optimal manufacturing process. Companies aspire to reduce this ratio as much as possible with the use of technology and maintenance tools.

Capacity Utilization

This production KPI measures the amount of capacity being utilized as a function of total capacity available. Ideally, companies want this number to be as high as possible, as it indicates they are making better use of their production capabilities and maximizing return on their assets. This metric can also be used by management when deciding whether to take on new orders or quoting lead time, as it gives a snapshot of available resources.

$$\text{Capacity Utilization} = \frac{Actual\ Factory\ Utilization}{Total\ Productive\ Capacity}$$

First Pass Yield

This is one of the most fundamental production's KPIs. It calculates the percentage of products manufactured to specification the first time through the process. This means that they do not require any rework or become scrap. A higher FPY rate is very desirable for any company.

$$\text{First Pass Yield Rate} = \frac{Quality\ Units}{Total\ Units\ Produced}$$

Predictive Production and Transparency in Delivery

Customer confidence increases greatly if order bookings, product preparation, and delivery are predictable and transparent. This is less of a production performance metric, but a very important KPI in the manufacturing sector. You can have the most efficient production line in the world, but if you can't deliver on time, clients are not going to want to work with you. This metric measures the percentage of products delivered to clients on time.

$$Ontime\ Delivery(\%) = \left(\frac{Number\ of\ Units\ Delivered\ Ontime}{Total\ Number\ of\ Units\ Delivered}\right) * 100$$

Both delivery to the client and your raw material intake should be automated, transparent, and predictable as part of an Intelligent Supply Chain as described in the next section. Some of the solutions to support in-time delivery are suggested below as part of a successful integrated ERP-MES implementation.

- Tracking of finished and semi-finished products on the shop floor throughout its shop-floor lifecycle
- Equipment Tracking Solutions for tracking tools, quality devices, portable machines and so on
- Based on Barcode reading, RFID, Ultra-Wideband, Bluetooth and 4G/5G technologies along with MES Integration to track material consumption

- Optimizing shop-floor warehouse inventory by interfacing with MES for data, combined with integrated SCM
- Facilitating integrated sales & operations, forecasting and demand, response and supply, and inventory planning by deploying an integrated ERP system

Intelligent Supply Chain

We think of the last 40 years as two eras of supply chain capability, each being 20 years. Companies started to adopt computer systems for production planning & procurement processes in the 1970s. This was a good first step towards value as it forced companies to structure foundational data like BOMs and move from reactive re-order point planning concepts to forward-looking demand-based production and procurement planning. This is where Material Requirements Planning (MRP) based Enterprise Resource Planning (ERP) systems were treated as the planning system of record.

However, it was quickly realized that their world had capacity constraints or material constraints from suppliers. Ignoring constraints on planning led to a lot of noise during execution and planner overload with MRP exceptions. This led to the second era of Advanced Planning Systems (APS). As shared by software solution provider Siemens on their webpage,

> *"Advanced Planning and Scheduling improves the synchronization of manufacturing processes and provides greater visibility to increase utilization and on-time delivery while reducing inventory levels and waste. Its advanced math enables manufacturers to quickly analyze*

and calculate achievable production schedules while considering multiple constraints and business rules. Planners can generate and evaluate what-if scenarios to achieve the best results."

Planners could model their entire supply network, forecast demand with statistical forecasting models, and model material and capacity constraints to produce optimal supply plans, which were sent back to ERP for execution. The planning system of records evolved and moved outside of ERP at this point. However, this mode of planning was still very enterprise-focused. Planners were still reliant on emails/spreadsheets to get feedback from key suppliers or sell-through demand trends from key customers / distributors. This led to latency in planning and value leakage. Also, internet and smart phones drove unprecedented levels of power to the end consumer, who could now research about products on the fly and switch brand loyalty if a better customer experience was offered elsewhere.

Speed is still the fundamental driving principle of supply chain management. It laid the foundation for the rise of modern-day successes like Dell, Walmart, FedEx, Amazon, etc. But for consumer-driven business models, success requires balancing velocity with another force that is important to growth – variety. For these businesses, increasingly, the message to the customer is "you are a unique individual with unique needs" where variety is central to the business strategy. And ground zero, where these two opposing forces of velocity and variety meet, is in manufacturing in the form of the Intelligent Supply Chain.

For instance, take the heading 'Digital Warehousing Improves Retailer Logistics by 40 Percent for Mexican Retail Company'. As Alfonso Núñez o (2022) wrote in his article Logistics & Mobility,

> *"Mexican retail companies have improved their logistics processes by up to 40 percent through the digitalization of their distribution centres and warehousing, according to data from the infrastructure solutions provider Panduit."*[43]

Real-time, data-driven supply chain, planning, and tracking of goods can give a competitive edge to success in many ways, as is detailed below:

- Automated Replenishment of Stock
- Precise Time of Arrival
- Production on Demand
- Route Optimization
- Warehouse Management – Reduced Inventory by 10%-50%
- Supply Network Simulation
- Reduced Starving of Production Lines – Higher productivity
- Increased Flexibility of Customized Production
- Reduced Plant Cycle Time
- Just-In-Time Delivery of Parts (JIT/JIS)
- Increased Capacity Utilization
- Intelligent Inventory Movement and Data Management model

In an Intelligent Supply Chain, you need to "Connect Digitally to Perfect Reality".

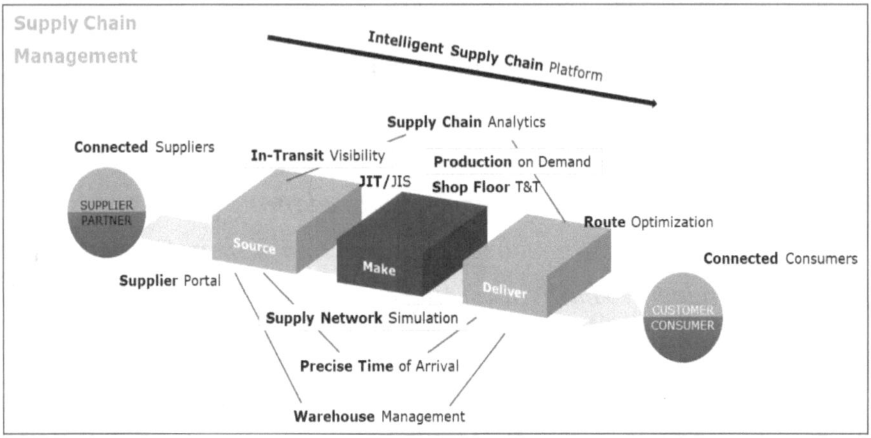

Figure 37: Intelligent Supply Chain Management solution

Connecting digitally is enabled through leveraging innovative technologies such as Blockchain, IOT, Big Data, Machine Learning, Blockchain, AI, Analytics and 3D printing.

IOT and unstructured data such as AI are generating previously unimaginable levels of Supply Chain Big Data. This Big Data is leveraged to drive smart assets and products with Machine Learning, and AI, and improve the end user experience and ability with Predictive Analytics. Blockchain can increase data security and trust with increased transparency, auditability, and regulatory compliance.

3D printing has great potential across the digital supply chain from rapid prototyping of designs, through additive manufacturing and in the rethinking of after-sales services and spare parts inventory through reduced inventory costs, increased flexibility and responsiveness and printing on demand services.

Perfecting reality requires the operational visibility and customer intelligence to manage your entire supply chain digitally and execute it flawlessly today, while transforming your business for the future.

Predictive and Prescriptive Analytics

Predictive analytics uses historical data, machine learning, and artificial intelligence to predict what will happen in the future. Predictive analytics can be applied to a common problem in the manufacturing industry. It helps to keep machines—and ultimately, the business itself—running.

Predictive Maintenance

Once you have trained a predictive model in the manufacturing space, you can use it to determine the likelihood of breakdowns. Plan ahead to shut machines down for preventive maintenance as needed to avoid unforeseen downtime or losses.

You can also use predictive analytics to limit or prevent any impact on your production planning. Also, by keeping an alert system in place on the cloud to monitor sensor's data (for example, motor vibration, temperature, or ambient temperature sensors), you can predict a failure.

Prescriptive Maintenance

Prescriptive Maintenance is another recipe to keep machines running without breakdown. Based on many years of experience of plant maintenance team and product experts, certain activities can be embedded into the system to suggest the changes to ensure longevity of machine operations. By knowing which machines will go down and when, the system can suggest and identify another machine or manufacturing shop that can pick up the missed load so your business doesn't slow down. For instance, avoid frequent start/stop of a motor if heated or using variable speed drive (VFD) or Soft-starter instead of a DoL starter for a motor in a long run if the motor is overheated frequently. All the algorithms and rules for predictive and prescriptive analytics can be fed into the system and matured over the years for smooth operations, improved availability of machines and continuous improvements in productivity.

Predictive Quality

The real-time actionable intelligence is delivered through AI & Machine Learning. The structured test data is continuously fed to machine learning algorithms that use techniques like data classification, regression, clustering, and association to predict product quality outcomes. The key performance indicators are reduced time to market and improved product quality. This, in turn, leads to customer satisfaction and major cost savings through effort optimization. As the AI engine matures through continuous learning, it delivers real-time insights that can improve test project outcomes on the go.

Predictive quality analytics aided by AI and machine learning can offer quality insights in seconds or minutes before the usefulness of these insights perish. It can also be compared or referenced with Golden-Batch Data to predict how close the product quality is to meeting the best quality attained so far within a given set of current process variables.

Additive Manufacturing or 3D printing

Imagine a steering knuckle for an electric vehicle, which could be produced with a weight reduction of more than 30% – thanks to 3D printing.

On the other hand, three-dimensional (3D) printing technologies have been in existence for roughly four decades, and, from the beginning, researchers and clinicians have been working to bring the technology into hospitals to improve patient care. 3D printing technology is currently impacting medical practice in almost every way imaginable, from high-profile cases like the separation of conjoined twins, to more day-to-day applications like surgical planning, patient counselling and student education. One of the most common hospital applications is the creation of patient-specific 3D-printed models of human anatomy, commonly referred to as point-of-care (POC) anatomical 3D printing. Created from a combination of MRI and CT scans of a real cadaver specimen, these models are able to print to life using cutting-edge 3D printers and materials from remote locations.

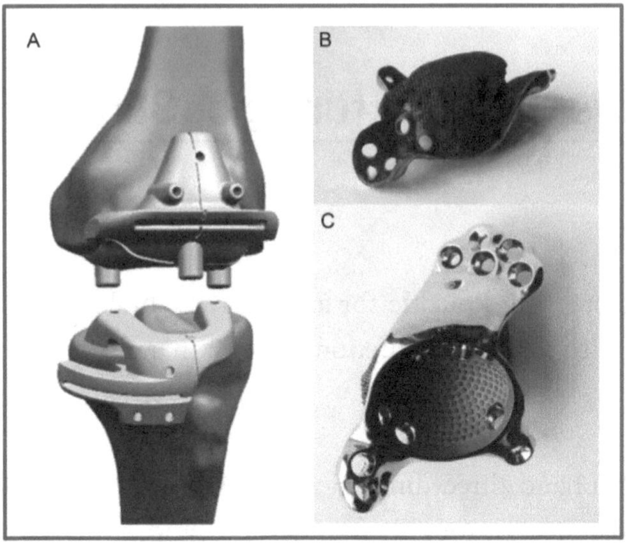

Figure 38: Potential of three-dimensional printing for orthopaedic applications[44]

Below is the flow chart of printing 3D parts (refers to an open Free Journal "From the printer: Potential of three-dimensional printing for orthopaedic applications")[45]

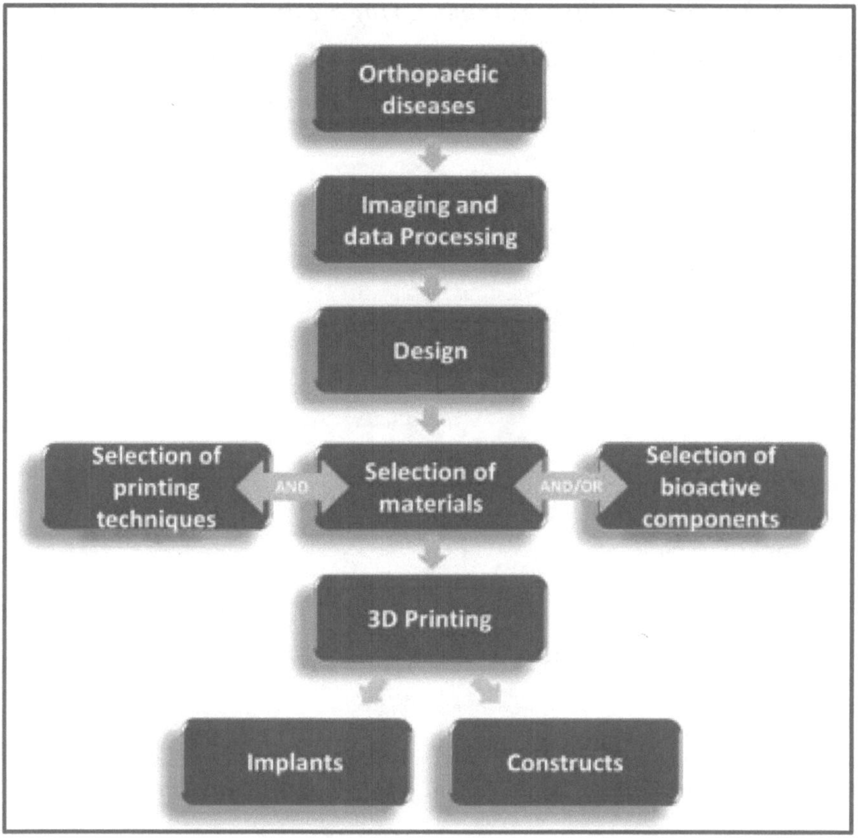

Figure 39: A Schematic Flow of creating three dimensional (3D) printed product[46]

Similarly industrial components or machine parts are being produced using 3D printers for multitude of benefits such as reduce time, improved quality, reliability and cost of production, etc. Melanie (2021) at Unleashed Software writes about benefits and disadvantages of additive manufacturing,

"The benefits of additive manufacturing are numerous, each of these individual benefits of additive manufacturing also have in common a certain degree of waste reduction and/or energy savings."[47]

Digital Twin

A digital twin is a virtual model designed to accurately reflect a physical object. The object being studied — for example, a wind turbine — is outfitted with various sensors related to vital areas of functionality. These sensors produce data about different aspects of the physical object's performance, such as energy output, temperature, weather conditions and more. This data is then relayed to a processing system and applied to the digital copy.

Once informed with such data, the virtual model can be used to run simulations, study performance issues and generate possible improvements, all with the goal of generating valuable insights — which can then be applied back to the original physical object.

The differences don't end there. For example, simulations usually don't benefit from having real-time data. However, digital twins are designed around a two-way flow of information that first occurs when object sensors provide relevant data to the system processor and then happens again when insights created by the processor are shared back with the original source object.

By having better and constantly updated data related to a wide range of areas, it can contextualize them by linking them in an efficient way such that it provides ample opportunity to compare performance differences between multiphysics models with artificial intelligence/ machine learning models. It can facilitate API-based secure data access to various simulation modelling tools for stress, thermal, fluid and electromagnetic issues and other simulations. It uses the feedback loop – it's able to provide field smart device, sensor and SCADA data back to the physics model to simulate real-world conditions on the digital twin's behaviour.

There are various types of digital twins depending on the level of product magnification. The biggest difference between these twins is the area of application. It is common to have different types of digital twins co-exist within a system or process. For instance, the availability of new IIoT-based solutions can be designed to optimize the Service Lifecycle Management (SLM) of assets by dynamically connecting real-world maintenance activities and asset performance back to design decisions and field modifications. The following are the types of digital twins that can be applied in different ways:

Component Twins/Parts Twins

Component twins are the basic unit of digital twin, the smallest example of a functioning component. Parts twins are roughly the same thing but pertain to components of slightly less importance.

Asset Twins

When two or more components work together, they form what is known as an asset. Asset twins let you study the interaction of those components, creating a wealth of performance data that can be processed and then turned into actionable insights.

System or Unit Twins

The next level of magnification involves system or unit twins, which enable you to see how different assets come together to form an entire functioning system. System twins provide visibility regarding the interaction of assets and may suggest performance enhancements.

Process Twins

Process twins, the macro level of magnification, reveal how systems work together to simulate an entire production facility. Are those systems all synchronized to operate at peak efficiency, or will delays in one system affect others? Process twins can help determine the precise timing schemes that ultimately influence overall effectiveness. Can you drive a car or operate a machine remotely using 5G technology?

Competitive Gains using Intelligent Manufacturing

Operational effectiveness may not necessarily be a strategy. Although you do something well, this doesn't signify it's strategically impactful for your business. You can offer a high-quality product or service, reduce inefficiencies in your business, and be more productive, but those things won't necessarily translate into driving more revenue or providing you with a sustainable competitive advantage in the marketplace.

Scheel, Henrik von (Dec 2019) replied to a question, "Is not adopting the Industry 4.0 paradigm already a problem for companies?"

> *"Yes, it is. If you are not adopting the changes, it basically disrupts your operations, business model, partnerships and your ability to stay relevant. And one of the main changes faced during adoption of Industry 4.0 is bridging the gap between strategy, design and delivery."*[48]

Businesses need both operational efficiency and strategy to create advantage and "superior performance." As Michael Porter, founder of the modern strategy field and one of the world's most influential thinkers on management and competitiveness, notes,

> *"A company can outperform rivals only if it can establish a difference that it can preserve. It must deliver greater value to customers or create comparable value at lower cost, or do both. Operational effectiveness means performing similar activities better than rivals perform them...In contrast, strategic positioning means performing different activities from rivals or performing similar activities in different ways."*[49]

Certainly, operational efficiency goes beyond cost management. It draws a strategy that looks both internally and externally at an organization's process for how they work. By understanding this strategy, you can identify areas where inefficiencies exist and determine how to improve them. Plus, by adopting Intelligent Manufacturing practices you can gain on operational efficiency, improve energy management and enhance the customer experience index by providing flexibility, transparency, online tracking, on-time delivery, superior product quality, customisation and so on. Igor Tymchuk, – Head of Data Unit, AVP Delivery of software development service firm "N-iX", noted (Oct 2020)[50],

> *"Leading manufacturers use Big Data to analyze the flow of finished goods and respond better to demand signals. IBM states that 53% of manufacturers apply Big Data and Data Analytics to create a competitive advantage for their organizations."*

Manufacturing companies need to develop a deep understanding of the technology's operational framework of translating business objectives into technology roadmaps targeting operational efficiency. This can be accomplished by leveraging the machine-to-machine communication, machine-to-human interaction, cloud computing and advanced analytics. For instance, intelligent machines can trigger maintenance processes autonomously and can predict failures; data analytics aid to detect process inefficiencies, thus reducing production cost. The Industry 4.0 ecosystem empowers manufacturers to plan long-term, sustainable business strategies, including leveraging cybersecurity as a business advantage. Talking about smoke detection, we can very well use sensors in cars, trucks or factories to monitor the toxic exhaust gas levels and take measures to minimise the damage to environment with help of IIoT technology.

In conclusion, on several accounts as stated above, Intelligent Manufacturing is a Smart Choice! The more we learn and start applying methodically it for betterment, more we gain.

Glossary

Acronym	Description
3D	Three-Dimensional
APS	Advanced Planning Systems
A&E	Alarms and Events
AI	Artificial Intelligence
AMQP	Advanced Message Queuing Protocol
API	Application Programming Interface
APM	Asset Performance Management
AWS	Amazon Web Services
B2B	Business to Business
B2C	Business to Consumer
BOM	Bill of Material
CNC	Computer Numerical Control
COM	Component Object Model
CIO	Chief Information Officer
CPU	Central Processing Unit
CRM	Customer Relationship Management
CTO	Chief Technology Officer
DA	Data Access
DB	Database
DCOM	Distributed Component Object Model

DoL	Direct On-Line starter for motors
ERP	Enterprise Resource Planning
ETL	Extract-Transform-Load of data
ICT	Information and Communications Technology, or Technologies
HA	Historical Access
HMI	Human Machine Interface
HTML	Hypertext Markup Language
HTTP	Hyper Text Transfer Protocol
HTTPS	Hyper Text Transfer Protocol -Secure
IAAS	Infrastructure as a Service
IEC	International Electrotechnical Commission
ISA	International Society of Automation
ISO	International Organization for Standardization
IIoT	Industrial Internet of Things
IoT	Internet of Things
IT	Information Technology system (IT system)
JDBC	Java Database Connectivity
MDM	Midsize Data Management
MES	Manufacturing Execution System
MESA	Manufacturing Enterprise Solution Association
MOM	Manufacturing Operations Management
ML	Machine Learning
MQTT	Message Queuing Telemetry Transport
MRP	Material Requirement Planning

NiFi S2S	NiFi Site-to-Site (NiFi was formerly called Niagarafiles)
OEE	Overall Equipment Effectiveness
OEM	Original Equipment Manufacturer
OOP	Object Oriented Programming
OPC	Open Platform Communication
OPC UA	Open Platform Communication – Unified Architecture
OPR	Operational Ratio
OSI	Open System Interconnection model
PAAS	Platform as a Service
PC	Personal Computer or Programmable Controller
PID	Proportional, Integral and Derivative control
PLC	Programmable Logic Controller
PLM	Product Lifecycle management
Pub-Sub	Publish-Subscribe
QMS	Quality Management System
RAMI 4.0	Reference Architectural Model for Industry 4.0
ROA	Return on Assets
ROI	Return On Investment
SaaS	"Software as a Service" on a cloud platform
SCADA	Supervisory Control and Data Acquisition system
SCM	Supply Chain management
SMS	Short Message Services
SOA	Service-oriented Architecture

SOAP	Simple Object Access Protocol
SRAM	Static Random-Access Memory
TCP/IP	Transmission Control protocol/ Internet Protocol
VFD	Variable Frequency Drive (or Variable speed drive)
VIN	Vehicle Identification Number
WMS	Warehouse Management System
ZVEI	German Electrical and Electronic Manufacturers' Association

Testimonials

My thoughts on this topic were well received by industry and professionals as highlighted below.

> Dear Sunil Wadhwa: "We take this opportunity to thank you on behalf of the entire team at Mahindra & Mahindra Ltd., FES for sharing your experience and expertise on the topic 'Advance Manufacturing Processes'.
>
> Your readiness to share your experience and expertise has played a vital role in making our event a success. You are an expert in your field and to have you as a speaker was the best we could do to inspire our employees. Your bright ideas and comforting guidance had a great positive impact on the attendees.
>
> Your talk has worked as a starting point to resolve important issues in the area and inspired many to take individual initiative.
>
> Once again, we thank you on behalf of everyone present at the conference for your kind gesture towards our company."

Figure 40: K, Sreenivasulu Manager HR, HRBP at Mahindra & Mahindra (Sep 2022) after a session conducted on "Advance Manufacturing" for 100+ their employees.51

"Dear Sunil Wadhwa, I would like to thank you for your valuable time for the interview for the topic of 'Impact of Agile Manufacturing on the Production Planning and Sequencing fo rAutomotive World.' Your expertise, your knowledge and command on this topic subject is remarkable and highly encouraging.

I have known you for a long time but never had a chance to have such a long conversation and it was great to learn a lot of new insights about agile execution, problems and challenges you face during execution in the real-world industry. The answers in in the interview questions would help me to analyse and formulate the survey questions towards my research on this topic in a great way.

Once again, thank you for your time and sharing your valuable knowledge. I wish all the best for your future endeavours in this field and for your writing in this area."

Figure 41: Hardikar, Saket (Sep 2022) wrote after conducting an interview on "Impact of Agile Manufacturing on production Planning and Sequencing on Automotive World.52

References

Books

Porter, Michael E., (30 Jun 2008) *The Competitive Strategy: Techniques for Analyzing Industries and Competitors* [online] Edition (Illustrated) Massachusetts https://www.amazon.co.uk/Competitive-Strategy-Techniques-Industries-Competitors-ebook/dp/B003YCQ8CA/ref=tmm_kin_swatch_0?_encoding=UTF8&qid=&sr=&asin=B003YCQ8CA&revisionId=&format=2&depth=1 (accessed on 30/10/2022)

Sutter, Burr (2017). *Teaching an elephant to dance e-book* [online] https://www.redhat.com/en/engage/teaching-an-elephant-to-dance-20180321?intcmp=701f2000000tjyaAAA&extIdCarryOver=true&sc_cid=70160000000cYRWAA2 (accessed on 30/10/2022)

Journals

Sze-Wing Mok , Razmara Nizak , Sai-Chuen Fu, Ki-Wai Kevin Ho , Ling Qin, Danie¨l B.F. Saris , Kai-Ming Chan, Jos Malda (2016) Open

Journal Journal of Orthopaedic Translation | ScienceDirect.com by Elsevier, P*atient-specific sawguides for total knee arthroplasty (Smith & Nephew, USA), "From the printer: Potential of three-dimensional printing for orthopaedic applications"*
www.sciencedirect.com (Accessed on Sep 2022)

Societies

International Society of Automation (Oct 2021) *Beyond the Pyramid: Using ISA95 for Industry 4.0 and Smart Manufacturing* [online] https://www.isa.org/intech-home/2021/october-2021/features/beyond-the-pyramid-using-isa95-for-industry-4-0-an (accessed on 21/09/2022)

MESA Foundation (1996) MESA-11 Model [online] https://mesa.org/topics-resources/mesa-model/history-of-the-mesa-models/ (accessed on 30/10/2022)

Publication IEC 62264-1:2013 (May 2013) ISO Enterprise-control system integration — Part 1: Models and terminology [online] https://www.iso.org/standard/57308.html, Switzerland, (sighted on 30/10/2022)

OPC Foundation [online] https://opcconnect.opcfoundation.org (accessed on 30/10/2022)

German Electrical and Electronic Manufacturers' Association [online] https://www.zvei.org/en/ (accessed on 30/10/2022)

Internet sources

Institutes and organizations

International Organization for Standardization https://www.iso.org/standard/57308.html

Endnotes

1 IDC (2021) IDC *FutureScape: Worldwide Digital Transformation 2022 Predictions* [online] https://www.idc.com/getdoc.jsp?containerId=US48312921 (accessed and downloaded on 30/10/2022)

2 Mims, Christopher (Apr 2015) WSJ *WSJ: Data Is the New Middle Manager* [online] https://www.wsj.com/articles/data-is-the-new-middle-manager-1429478017

3 Littlefield, Matthew (2016) Whitepaper *IIoT AND BIG DATA ANALYTICS: How Manufacturing System Architecture Is Being Transformed* LNS Research. [Online] www.lnsresearch.com (accessed and downloaded on 18/07/2018)

4 Kelly, Tom (2019), Automation Alley USA *Industry 4.0: From Vision to Implementation* [Online] https://www.ptc.com/-/media/Files/PDFs/IoT/Automation-Alley_Technology-in-Industry-Report.pdf?sc_lang=en (download and accessed Oct 2022)

5 Scheel, Henrik Von (2020) *Who is Henrik Von Scheel–Leading authority on strategy and competitiveness* [Online] http://von-scheel.com/biography/ (Accessed on 25/10/2022)

6 Williamson, Jonny (2019) *Industry 4.0 arrives in waves that disrupt everything. Manufacturing Leaders' Summit: Henrik Von Scheel on moving beyond the hype of Industry 4.0 – The Manufacturer.* [Online] https://www.themanufacturer.com/articles/manufacturing-leaders-summit-henrik-von-scheel-on-moving-beyond-the-hype-of-industry-4-0/ (Accessed on 25/10/2022)

7 MacDougall, William (July 2014) Germany Trade and Invest (GTAI) industrie4-0-smart-manufacturing-for-the-future-en.pdf (Germany) [online] https://www.pac.gr/bcm/uploads/industrie4-0-smart-manufacturing-for-the-future-en.pdf (Downloaded on 24/10/2022)

8 Willamson, J. (2019) *Manufacturing Leaders' Summit: Henrik Von Scheel on moving beyond the hype of Industry 4.0, UK*, The Manufacturer [online]. Manufacturing Leaders' Summit: Henrik Von Scheel on moving beyond the hype of Industry 4.0 - The Manufacturer (Accessed on 21 August 2022)

9 MacDougall, William (2018) *industrie4-0-germany-market-outlook-progress-report-en-data* Germany [online] https://www.gtai.de/en/invest/service/publications/industrie-4-0-germany-market-report-and-outlook-64602 (Accessed on 21 August 2022)

10 Mortson Enterprises Inc. (2017) *Supply Chain Game Changer* [Online] https://supplychaingamechanger.com/industry-4-0-ecosystem-infographic/ (accessed on 30/10/2022)

11 Whitepaper released by Orange (Dec 2018), *Orange_5G_iiot_whitepaper_dec2018* [online] https://www.orange-business.com/en/library/whitepaper/industrial-iot (accessed and download on 28/09/2022)

12 Tymchuk, Igor (Oct 2020) Head of Data Unit, AVP Delivery at "N-iX" *Big Data in manufacturing made simple: Trends, tips, success stories* [online]. https://www.n-ix.com/big-data-in-manufacturing/ (accessed on 03/10/2022)

13 Financial Times (2012) *Why Big Data is a Forth Factor of production* [online] https://www.ft.com/content/5086d700-504a-11e2-9b66-00144feab49a (accessed on 03/10/2022)

14 Russom, Philip (2011) TWDI Research *The Three Vs of Big Data Analytics* [online] TDWI.org https://tdwi.org/blogs/tdwi-blog/2011/06/three-vs-of-big-data-analytics-1-data-volume.aspx (accessed on 03/10/2022)

15 Samuel, Nicholas (October, 2021) *Data Integration, Data Lake, Data Lake Architecture: 10 Critical Aspects* [online] https://hevodata.com/learn/data-lake-architecture-a-comprehensive-guide/#datalakearchitecture (accessed on 11-09-2022)

16 Samuel, Nicholas (October, 2021) *Data Integration, Data Lake, Data Lake Architecture: 10 Critical Aspects* [online] https://hevodata.com/learn/data-lake-architecture-a-comprehensive-guide/#datalakearchitecture (accessed on 11-09-2022)

17 Samuel, Nicholas (October, 2021) *Data Integration, Data Lake, Data Lake Architecture: 10 Critical Aspects* [online] https://hevodata.com/learn/data-lake-architecture-a-comprehensive-guide/#datalakearchitecture (accessed on 11-09-2022)

18 Stepanenko, Dmitriy (Feb 2019) Distillery *Data Warehouse vs. Data Lake: Which Is Right for Your Enterprise App Development Effort?* [online] https://distillery.com/blog/data-warehouse-vs-data-lake-which-is-right-for-your-enterprise-app-development-effort/#:~:text=Data%20Lake%20vs.&text=A%20data%20lake%20is%20often,entire%20scope%20of%20the%20business (accessed on 11-09-2022)

19 Stepanenko, Dmitriy (Feb 2019) Distillery *Data Warehouse vs. Data Lake: Which Is Right for Your Enterprise App Development Effort?* [online] https://distillery.com/blog/data-warehouse-vs-data-lake-which-is-right-for-your-enterprise-app-development-effort/#:~:text=Data%20Lake%20vs.&text=A%20data%20lake%20is%20often,entire%20scope%20of%20the%20business (accessed on 11-09-2022)

20 Stepanenko, Dmitriy (Feb 2019) Distillery *Data Warehouse vs. Data Lake: Which Is Right for Your Enterprise App Development Effort?* [online] https://distillery.com/blog/data-warehouse-vs-data-lake-which-is-right-for-your-enterprise-app-development-effort/#:~:text=Data%20Lake%20vs.&text=A%20data%20lake%20is%20often,entire%20scope%20of%20the%20business (accessed on 11-09-2022)

21 International Society of Automation (Oct 2021) *Beyond the Pyramid: Using ISA95 for Industry 4.0 and Smart Manufacturing* [online] https://www.isa.org/

intech-home/2021/october-2021/features/beyond-the-pyramid-using-isa95-for-industry-4-0-an (accessed on 21/09/2022)

22 Gills, Alexander S. (July 2020) Technical Writer and Editor *enterprise architecture* on TechTarget webpage [online] https://www.techtarget.com/searchcio/definition/enterprise-architecture (accessed on 21/09/2022)

23 SAP SE (Mar 2016) *Reshape the Automation Pyramid* [online] https://opcconnect.opcfoundation.org/2016/03/reshape-the-automation-pyramid/ (accessed on 22/09/2022)

24 SAP SE (Mar 2016) *Reshape the Automation Pyramid* [online] https://opcconnect.opcfoundation.org/2016/03/reshape-the-automation-pyramid/ (accessed on 22/09/2022)

25 Hankel, Martin and Rexroth, Bosch (Mar 2015) *RAMI4.0 Map of Reference Architecture Model for Industry 4.0* [online] https://www.zvei.org/en/ (sighted and downloaded on 12/3/2022)

26 Hankel, Martin and Rexroth, Bosch (Mar 2015) *RAMI4.0 Map of Reference Architecture Model for Industry 4.0* [online] https://www.zvei.org/en/ (sighted and downloaded on 12/3/2022)

27 Publication IEC 62264-1:2013 (May 2013) *ISO Enterprise-control system integration — Part 1: Models and terminology* [online] https://www.iso.org/standard/57308.html, Switzerland, (sighted on 12/3/2022)

28 Sharma, Priyanka (2021) GM at Cloud Native Computing Foundation *Building Sustainable Ecosystems for Cloud Native Software* [online] https://www.cncf.io/ (accessed on 30/10/2022)

29 Hofmann, Martin Dr. (2019) Volkswagen AWS re:Invent 2019 – Connects 124 *Factories to the Cloud Using AWS|Volkswagen Group Cloud* [online] https://aws.amazon.com/solutions/case-studies/vw-industrial-cloud/ (sighted on 30/10/2022)

30 Amazon Press Release (March 27, 2019) *Volkswagen and AWS Join Forces to Transform Automotive Manufacturing* [online] Volkswagen and AWS Join Forces to Transform Automotive Manufacturing | Amazon.com, Inc. - Press Room (aboutamazon.com) https://press.aboutamazon.com/news-releases/news-release-details/volkswagen-and-aws-join-forces-transform-automotive (accessed on 12-3-2022)

31 Interaction Design Foundation, CC BY-SA 3.0 [online] *What is Design Thinking?* | [online] https://www.interaction-design.org/literature/topics/design-thinking (Accessed on 31/08/2022)

32 **inray Industriesoftware GmbH** *OPC Router The OPC UA Pub/Sub Plug-in* [online] https://www.opc-router.com/ (accessed on 22/09/2022)

33 Lutz, Peter (Sep 2021) "*FLC Corner – September 2021* [online] OPCFoundation https://opcconnect.opcfoundation.org/2021/09/flc-corner-september-2021/ (accessed on 22/09/2022)

34 OPC Foundation's online reference publication (2022) P*ubSub Concepts. OPC UA* [online] https://reference.opcfoundation.org/v104/Core/docs/Part14/5.1/ (Accessed on 31/08/2022)

35 OPC Foundation's online reference publication (2022) P*ubSub Concepts. OPC UA* [online] https://reference.opcfoundation.org/v104/Core/docs/Part14/5.1/ (Accessed on 31/08/2022)

36 Brebnr, Paul (May 2018) *Apache Kafka® Connect Architecture Overview* [online] https://www.instaclustr.com/blog/apache-kafka-connect-architecture-overview/ (accessed on 23/09/2022)

37 MESA Foundation (1996) MESA-11 Model [online] https://mesa.org/topics-resources/mesa-model/history-of-the-mesa-models/ (accessed on 30/10/2022)

38 The Verge – Automated Bots at Ocado Andover Story https://www.theverge.com/2018/5/8/17331250/automated-warehouses-jobs-ocado-andover-amazon

39 Siemens Industry Support – Demo Projects [online]. Entry on 02/03/2022. https://support.industry.siemens.com/cs/ww/en/view/109806181 (Accessed on 03/10/2022)

Sze-Wing Mok , Razmara Nizak , Sai-Chuen Fu, Ki-Wai Kevin Ho , Ling Qin, Danie¨l B.F. Saris , Kai-Ming Chan, Jos Malda (2016) *Open Journal Journal of Orthopaedic Translation* | ScienceDirect.com *by Elsevier*, Patient-specific sawguides for total knee arthroplasty (Smith & Nephew, USA), "From the printer: Potential of three-dimensional printing for orthopaedic applications" www.sciencedirect.com (Accessed on Sep 2022)

40 Sutter, Burr (2017) *Teaching an elephant to dance e-book* [online] https://www.redhat.com/en/engage/teaching-an-elephant-to-dance-20180321?intcmp=701f2000000tjyaAAA&extIdCarryOver=true&sc_cid=70160000000cYRWAA2 (Accessed on 8/01/2020)

41 International Society of Automation (2021) *Beyond the Pyramid: Using ISA95 for Industry 4.0 and Smart Manufacturing* [online] https://www.isa.org/intech-home/2021/october-2021/features/beyond-the-pyramid-using-isa95-for-industry-4-0-an (accessed on 21/09/2022)

42 Bush, Stephan (2022). *The Effects of Working Capital in the Operation of a Manufacturing Industry* newsletter CHRON [online] https://smallbusiness.chron.com/effects-working-capital-operation-manufacturing-industry-78120.html (accessed on 24/09/2022)

43 Núñez, Alfonso, (2022) *Logistics & Mobility* https://mexicobusiness.news/mobility/news/digital-warehousing-improves-retailer-logistics-40-percent (accessed on 25/09/2022)

44 Sze-Wing Mok , Razmara Nizak , Sai-Chuen Fu, Ki-Wai Kevin Ho , Ling Qin, Danie¨l B.F. Saris , Kai-Ming Chan, Jos Malda (2016) Open Journal *Journal of Orthopaedic Translation | ScienceDirect.com by Elsevier, Patient-specific sawguides for total knee arthroplasty (Smith & Nephew, USA), "From the printer: Potential of three-dimensional printing for orthopaedic applications"* www.sciencedirect.com (Accessed on Sep 2022)

45 Sze-Wing Mok , Razmara Nizak , Sai-Chuen Fu, Ki-Wai Kevin Ho , Ling Qin, Danie¨l B.F. Saris , Kai-Ming Chan, Jos Malda (2016) Open Journal *Journal of Orthopaedic Translation* | ScienceDirect.com *by Elsevier*, Patient-specific sawguides for total knee arthroplasty (Smith & Nephew, USA), "From the printer: Potential of three-dimensional printing for orthopaedic applications" www.sciencedirect.com (Accessed on Sep 2022)

46 Sze-Wing Mok , Razmara Nizak , Sai-Chuen Fu, Ki-Wai Kevin Ho , Ling Qin, Danie¨l B.F. Saris , Kai-Ming Chan, Jos Malda (2016) Open Journal *Journal of Orthopaedic Translation | ScienceDirect.com by Elsevier, Patient-specific sawguides for total knee arthroplasty (Smith & Nephew, USA), "From the printer: Potential of three-dimensional printing for orthopaedic applications"* www.sciencedirect.com (accessed on 25/09/2022)

47 Melanie (2012) *12 Benefits of Additive Manufacturing and 5 Disadvantages* [online] https://www.unleashedsoftware.com/blog/12-benefits-of-additive-manufacturing-and-5-disadvantages?utm_source=adwords&utm_term=&utm_campaign=%5BG%5D+DSA+-+UK&utm_medium=ppc&hsa_ver=3&hsa_kw=&hsa_src=g&hsa_cam=1039441623&hsa_net=adwords&hsa_acc=5012851927&hsa_mt=&hsa_grp=50591493563&hsa_tgt=dsa-19959388920&hsa_ad=245745873260&gclid=Cj0KCQjwnvOaBhDTARIsAJf8eVOSaa6hKAGPTi3eQoSvzL3Xc_FW_fQC_WudIV1ThRgiupaoPSoDYqoaApZ-EALw_wcB (accessed on 29/10/2022)

48 García, Juan Francisco (2019) *Digital Enterprise Show 2019* by Digital Biz Magazine [Online] https://www.digitalbizmagazine.com/henrik-von-scheel/ (Accessed on 25/10/022)

49 Porter, Michael E., (30 Jun 2008) *The Competitive Strategy: Techniques for Analyzing Industries and Competitors* [online] Edition (Illustrated) Massachusetts: Free Press. Available from https://www.amazon.co.uk/Competitive-Strategy-Techniques-Industries-Competitors-ebook/dp/B003YCQ8CA/ref=tmm_kin_swatch_0?_encoding=UTF8&qid=&sr=&asin=B003YCQ-8CA&revisionId=&format=2&depth=1 (accessed on 29/10/2022)

50 Tymchuk, Igor – Head of Data Unit, AVP Delivery of software development service firm "N-iX" (Oct 2020). Big Data in manufacturing made simple: Trends, tips, success stories [online]. Available on https://www.n-ix.com/big-data-in-manufacturing/ (accessed on 03/10/2022)

51 K, Sreevilasulu Manager HR, HRBP at Mahindra & Mahindra (Sep 2022) after a session conducted on "Advance Manufacturing" for 100+ their employees. https://www.linkedin.com/posts/activity-6980735902310379520-rnqN?utm_source=share&utm_medium=member_android (accessed on 25/09/2022)

52 Hardikar, Saket, (Sep 2022) interview debrief on *Impact of Agile Manufacturing on Production Planning and Sequencing for Automotive World* https://www.linkedin.com/posts/sakethardikar_manufacturing-thankyou-automotive-activity-6981628599183265793-uBO-?utm_source=share&utm_medium=member_android (accessed on 25/09/2022)

Made in United States
Troutdale, OR
11/03/2026

56845269R00106